Uncovering Submerged Landscapes

Towards a GIS method for locating submerged archaeology in southeast Alaska

Kelly Rose Bale Monteleone

BAR INTERNATIONAL SERIES 2917 | 2019

Published in 2019 by
BAR Publishing, Oxford

BAR International Series 2917

Uncovering Submerged Landscapes

ISBN 978 1 4073 1656 7

© Kelly Rose Bale Monteleone 2019

COVER IMAGE *Sunset in Shipley Bay, June 2010.*

The Author's moral rights under the 1988 UK Copyright,
Designs and Patents Act are hereby expressly asserted.

All rights reserved. No part of this work may be copied, reproduced, stored,
sold, distributed, scanned, saved in any form of digital format or transmitted
in any form digitally, without the written permission of the Publisher.

Printed in England

BAR titles are available from:

BAR Publishing
122 Banbury Rd, Oxford, OX2 7BP, UK
EMAIL info@barpublishing.com
PHONE +44 (0)1865 310431
FAX   +44 (0)1865 316916
www.barpublishing.com

Acknowledgements

I would like to acknowledge and thank all of my dissertation committee members for their advice, guidance, and patience. They were Dr E. James Dixon, Dr Michael W. Graves, Dr David Dinwoodie, Dr Karl Benedict, and Dr Grant Meyer. In particular, I am grateful for the support, encouragement, and patience of my dissertation advisor, Dr Dixon. His focus and steady encouragement helped me complete my PhD in a timely manner and encouraged me to continue this research.

I would like to acknowledge my research team from the University of New Mexico. I would also like to thank the many friends who read portions of the research, talked me through the more difficult parts and were always there to cheer me on. Finally, Jason Brown at EDAC (Earth Data Analysis Center at UNM) for his help writing Python Scripts and Dr Andrew Wickert from the University of Minnesota for all of the past and continuing research collaboration.

I would like to acknowledge my parents and family. My mom, Claudia Monteleone has supported me emotionally through this journey. I would like to thank my brother and sister (Adam and Lindsey Monteleone), father (Frank Monteleone), and stepmother (Lorna Monteleone) who passed away last year, for their support and encouragement through this long and ongoing research.

Finally, I would like to thank the Office of Polar Programs at the National Science Foundation for the funding for this research; Tim Bulman at InDepth Marine who provided support when no one else could; my supervisor and colleagues in enrolment services at Mount Royal University who have dealt-with the distributions to my day-job to complete this publication.

Contents

List of Tables xi
List of Figures viii
Abstract xii

1 Introduction 1
    1.1. Introduction 1
    1.2. Project Background 1
    1.3. Study Region and Study Area 3
    1.4. The GIS Model 3
    1.5. Peopling of the Americas 4
    1.6. Northwest Coast Maritime Adaptations 4
    1.7. How Does This Relate to Other Projects? 5
    1.8. Structure of This Book 5

2 Theoretical Background 7
    2.1. Introduction 7
    2.2. Landscape Theory 7
        2.2.1. Historical Development of Landscape Theory 7
        2.2.2. Modern Landscape Theory 8
        2.2.3. Non-sites 10
        2.2.4. Seascapes 10
    2.3. Archaeological Predictive Models 11
        2.3.1. History of High Potential Models 11
        2.3.2. High Potential Models as Middle-Range Theory 12
        2.3.3. Types of High Potential Models 12
        2.3.4. Testing Models 13
        2.3.5. Issues with Predictive Models 14
            2.3.5.1. Site Types 14
            2.3.5.2. Archaeological Site Sample 15
            2.3.5.3. Palaeoenvironments 15
            2.3.5.4. Scale 16
        2.3.6. Examples of Predictive Models 17
    2.4. Geographic Information Systems (GIS) 17
        2.4.1. History of GIS 18
        2.4.2. GIS Programs 18
    2.5. Underwater Archaeology 18
        2.5.1. Danish Model 19
        2.5.2. Underwater Archaeological Projects 20
    2.6. Conclusion 22

3 Introduction to the Pre–10,000 cal BP Palaeogeography of the northern NWC and Vicinity 23
    3.1. Introduction 23
    3.2. Southeast Alaska 23
    3.3. Flora and Fauna 24
    3.4. Glaciation 25
    3.5. Sea Level Reconstructions 26
        3.5.1. Sea Level Reconstruction for the Outer Islands of the Alexander Archipelago 30
        3.5.2. Sea Level Reconstruction for the Inner Islands of the Alexander Archipelago 30
    3.6. Last Glacial Maximum Refugia in Southeast Alaska 30
        3.6.1. Faunal Evidence Supporting Refugia 30
        3.6.2. Flora Evidence Supporting Refugia 31
    3.7. Geomorphology of Southeast Alaska 33

  3.8. Study Area – Shakan Bay 34

4 Archaeology and Ethnography of the northern NWC 37
 4.1. Introduction 37
 4.2. Origins and Land-Use History of the NWC 37
 4.3. Cultural Chronologies 41
  4.3.1. Variability in Site Locations 42
 4.4. Geographic Variability 42
 4.5. Northern NWC Ethnographic Groups 44
  4.5.1. Tsimshian 44
  4.5.2. Haida 46
  4.5.3. Tlingit 47
 4.6. Site Location Descriptions 48

5 The GIS Model 51
 5.1. Introduction 51
 5.2. Data 51
  5.2.1. Bathymetry and DEM 51
  5.2.2. Archaeological Site Data 56
 5.3. The Three Stages of the Model 58
  5.3.1. How This Research Fits with Similar Site Discovery Models 60
  5.3.2. Stage One – Inputs for the Model 61
   5.3.2.1. Water 62
  5.3.3. Stage Two – Model Outputs 62
   5.3.3.1. Intermediate Products 62
   5.3.3.2. Coasts and Coastal Sinuosity 65
   5.3.3.3. Final Products 67
 5.4. Summary 68

6 Analysis of the Model 69
 6.1. Introduction 69
 6.2. Kvamme's Gain 69
 6.3. Spatial Statistics 70
  6.3.1. Getis-Ord General-G 70
  6.3.2. Global Moran's I 71
 6.4. Summary 72

7 Field Testing the Model 75
 7.1. Introduction 75
 7.2. June 2010 Survey 75
  7.2.1. Shakan Bay Anomaly One 77
  7.2.2. Shakan Bay Anomaly Two 78
  7.2.3. Shakan Bay Anomaly Three 78
  7.2.4. Shakan Bay Anomaly Five - Shipwreck 80
 7.3. May 2012 Survey 80
  7.3.1. Multibeam Sonar 81
  7.3.2. Sub-bottom Profiler 83
  7.3.3. ROV and Sub-surface Testing 84
 7.4. Summary 86

8 Pre-10,000 cal BP Archaeological Sites on the Northern NWC 87
 8.1. Introduction 87
 8.2. Southeast Alaskan Sites 87
  8.2.1. Ground Hog Bay 2 (11,528 cal BP) 87
  8.2.2. Hidden Falls (10,157 cal BP) 89
  8.2.3. Irish Creek (10,484 cal BP) 89
  8.2.4. On Your Knees Cave (10,207 cal BP and 12,129 cal BP) 90
  8.2.5. Trout Creek Upper Terrace (10,364 cal BP) 90
  8.2.6. Rice Creek (10,235 cal BP) 91

- 8.3. Haida Gwaii 91
  - 8.3.1. K1 Cave (12,650 cal BP) 91
  - 8.3.2. Richardson Island (10,442 cal BP) 91
  - 8.3.3. Werner Bay (12,481 cal BP) 92
  - 8.3.4. Arrow Creek (10,625 cal BP and 10,584 cal BP) 92
  - 8.3.5. Gaadu Din Caves (12,683 cal BP cave 1, 12,480 cal BP cave 2) 92
  - 8.3.6. Kilgii Gwaay (10,675 cal BP) 93
- 8.4. Mainland British Columbia 93
  - 8.4.1. Far West Point (11,045 cal BP) 93
  - 8.4.2. Triquet Island (13,850 cal BP) 94
  - 8.4.3. Kildidit Narrows (11,419 cal BP and 13,564 cal BP) 94
  - 8.4.4. Namu (11,049 cal BP) 94
  - 8.4.5. North Pruth Bay (10,608 cal BP) 95
  - 8.4.6. North Kwakshua Channel (13,300 cal BP) 95
- 8.5. Site Summary 95

9 Discussion
- 9.1. Introduction 97
- 9.2. Model 97
  - 9.2.1. Resolution 97
  - 9.2.2. Moderately High vs. High Potential 97
  - 9.2.3. Spatial Statistics 98
- 9.3. Shakan Bay Glacial History 98
- 9.4. Why Were No Sites Located? 98
- 9.5. Beringia 99

10 Conclusion 101

Bibliography 103

List of Figures

Figure 1-1. Map of study region and area 2

Figure 3-1. Map of the geographic region 24

Figure 3-2. Map of Pleistocene glaciation with the NWC enlarged 25

Figure 3-3. LGM glaciation and possible refugia 27

Figure 3-4. Sea level reconstructions for the NWC 28

Figure 3-5. Changes in land-sea relation. A) Depicts the elastic or instantaneous movement of the land-sea relationship with glacial melt; here the land is rising due to isostatic rebound as the glaciers melt; the sea is rising due to eustatic sea level rise, with a small locally lower sea level near the land/glacier due to gravity. B) Depicts the post-glacial response; the sea level is lowering and the land where the glaciers once were rising due to isostatic rebound. The forebudge that is now under the sea as it collapses or lowers. C) Depicts the LGM approximately 24,000 to 17,000 cal BP when sea level was 165 meters lower than modern level. The reconstructed glacial extent is similar to 1600 meters ELA. D) Depicts an early post-glacial reconstruction from between approximately 15,000 to 11,000 cal BP where sea level was 70 meters below modern levels. The reconstructed glacial extent is similar to 2000 meters ELA. Here land that was deglaciated rises due to isostatic rebound. Sea level rises due to global eustatic sea level rise. E) Depicts the early Holocene from approximately 10,500 to 5500 cal BP where sea level is up to 20 meters above modern levels with modern glacial extent; the ground lowers or settles after the isostatic rebound to near modern levels. Sea levels continue to rise. Towards the mainland, the land is uplifted due to isostatic rebound. F) Depicts the late Holocene with modern sea level, glacial extent, and land surface; the major island of the cross section are labelled for reference (POWI = Prince of Wales Island). The thin black jagged line is the cross section through Shakan Bay associated with the modern sea level, or thin black horizontal line 29

Figure 3-6. Flora pollen sequences 32

Figure 3-7. Coastal zone from the coastal plane to the continental shelf 33

Figure 3-8. Average wave period from Cape Edgecumbe Buoy (46084) 34

Figure 3-9. Map of Shakan and Shipley Bays depicting the geology, the bathymetry, and locations named in text 35

Figure 4-1. Northwest Coast including locations discussed in chapter 38

Figure 4-2. Published chronologies of the NWC 39

Figure 4-3. NWC language families 40

Figure 4-4. Northern NWC ethnographic groups 45

Figure 5-1. Distribution of data points included in the bathymetry DEM of the study area 52

Figure 5-2. DEM generated for this project over NOAA chart with matching contour lines (black lines NOAA, red lines DEM) 53

Figure 5-3. Comparison of Spline and IDW DEM methods. A) Spline DEM. B) IDW DEM. C) The absolute value of IDW minus Spline DEMs. The comparison was conducted between the IDW and a Spline version of the DEM to determine if there would be a significant difference in results between the IDW and a Spline version of the DEM 54

List of Figures

Figure 5-4. Multibeam. A) Multibeam data collected for Shakan Bay in 2012. B) Difference between DEM and multibeam at 1.5 m resolution 55

Figure 5-5. Map of archaeological sites in the project database (see table 5-1 for site types) 56

Figure 5-6. Flowchart of three stages for developing the model 57

Figure 5-7. Lake and streams reconstructed in Shakan Bay study area for 16,000 cal BP, darker land is modern 58

Figure 5-8. Reclassified slope raster for Shakan Bay study area 16,000 cal BP (two-meter resolution) 60

Figure 5-9. Reclassified aspect raster for Shakan Bay study area 16,000 cal BP (two-meter resolution) 61

Figure 5-10. Histogram of distance to features (lakes, rivers, tributary junctions, coastline, and other sites) from known archaeological site location 63

Figure 5-11. Histogram of distance to features (lakes, rivers, tributary junctions, coastline, and other sites) from random locations 63

Figure 5-12. Sinuosity is the average of the length along the coastline divided by the linear length of the line segment within a three-kilometre buffer diameter around each point along the coast 64

Figure 5-13. Coastal sinuosity of Shakan Bay Area at 13,000 cal BP including high, medium, and low rank for archaeological sites based on statistical analysis 65

Figure 5-14. Stacked image of rasters used in final weighted model (13,000 cal BP) 66

Figure 5-15. Model results. A) Map of Shakan Bay at 13,000 cal BP as an example of a time slice of the final model. B) Map of Shakan Bay with the model results showing the final high potential model with model values from 11,000 to 16,000 cal BP combined. C) Histogram of 13,000 cal BP model based on values. D) Histogram of final model results. The histograms show the full values of the weighted overlay. High potential is value four and moderately high archaeological potential is value three 67

Figure 7-1. Map of Shakan Bay side scan survey from June 2010 indicating anomaly locations. The modern land extent is in grey with model the background 75

Figure 7-2. Shakan Bay Anomaly one: two circular and one rectangular depression (white space). Inset has depressions outlined in dotted lines 76

Figure 7-3. Shakan Bay anomaly two: two raised rectangular features (darker grey) 77

Figure 7-4. Shakan Anomaly three: raised semi-circular features and two depressions. Inset has raised features and circular depressions indicated with dotted lines.. 78

Figure 7-5. Reconstruction of the location of the potential weir and depressions on the landscape (sea level is at 52 m with a one-meter gap for tide)... 79

Figure 7-6. Shakan anomaly five: shipwreck (PET-714), likely the Restless lost in 1910. Inset has raised features outline with a dotted line 79

Figure 7-7. Density of multibeam points per 1m². Gaps in multibeam coverage are clearly visible as white space 80

Figure 7-8. Multibeam results at 1.5-m resolution (with five-meter colour palate). The multibeam data extends to the east, where the *R/V Antonie* harboured each night. Data were collected as we approached the survey area 81

Figure 7-9. Example of the poor quality of sub-bottom data due to calibration issues 82

Figure 7-10. Map of identified sub-bottom anomalies. When 'a' and 'b' are included, the anomaly stretches over a large area and the two locations mark the beginning and end of the anomaly or feature, such as figure 7-14 82

Figure 7-11. Sub-bottom image of a possible filled stream from location A26 83

Figure 7-12. Sub-bottom image of a possible filled depression from location A15 83

Figure 7-13. Sub-bottom image includes a discontinuous recorded surface with buried sediment layer from location A5 84

Figure 7-14. Sub-bottom image of possible shell deposit from location A9 84

Figure 7-15. Van veen grab sampler used for sub-surface testing (Leatherman used as scale) 85

Figure 7-16. Map of ROV investigations or dives and sub-surface test locations 85

Figure 7-17. Cut wood from VV25-20 from Shakan Bay Area three 86

Figure 7-18. Piece of rounded wood recovered from Area seven (VV-26-06) 86

Figure 8-1. Map of Pre-10,000 cal BP archaeological sites from the northern NWC 88

Figure 9-1. Map of Bering Sea with inset map of southern Beringian archipelago 100

List of Tables

Table 3-1. Tidal range benchmarks around POWI in meters relative to mean lower low water (MLLW) 34

Table 4-1. Results of grouped ANOVA for study region; the null hypothesis can be accepted, or that the chronologies are statistically the same except for slope (in bold) as it has a p-value of less than alpha 43

Table 4-2. Results of paired t-tests for slope where the null hypothesis can be rejected 43

Table 5-1. Frequency of archaeological site types within the entire database, study region, and study area. New sites are all in the study area and were not part of the model production 57

Table 5-2. Results of Welsh two-sample t-test for model variables between archaeological site locations and random locations 59

Table 5-3. Coastal sinuosity ranks in relation to the location of recorded sites 61

Table 5-4. Weighted overlay variables, ranking, and weights for the model 62

Table 6-1. Kvamme's Gain statistics 69

Table 6-2. Getis-Ord General-G and Global Moran's I statistics using inverse-distance 71

Table 6-3. Global Moran's I results using fixed distance bands 72

Table 8-1. Average calibrated 14C dates for sites older than 10,000 cal BP 87

Abstract

Early peoples migrating to the Americas using the coastal migration route would have travelled through southeast Alaska during lower sea levels. The residues of where they lived, hunted, and gathered are on the now submerged continental shelf of southeast Alaska. This project tests the hypothesis that the archaeological record of southeast Alaska extends to areas of the continental shelf that were submerged by post-Pleistocene sea level rise prior to 10,600 cal years BP (9,400 ^{14}C years). The study region is the southern outer island of the Alexandra Archipelago in southeast Alaska; an area occupied by the modern Tlingit, Haida, and Tsimshian peoples. This region has areas that remained unglaciated during the last glacial maximum where bears, deermice, and other mammals survived in pockets of refugium. When the last glacial maximum ended locally between 19,000 and 16,000 cal BP, this region quickly recovered with plant and animal life, because of the refugium. The current earliest evidence of people on the northern Northwest Coast of the Americas is at Triquet Island near Namu, British Columbia where a hearth has been dated to 13,300 cal BP. Earlier and contemporaneous evidence of people on the Northwest Coast is anticipated to be submerged on the continental shelf. This is anticipated because, following Erlandson's kelp highway theory for the peopling of the Americas, people within the northwest coast were maritime adapted when they migrated to the Americas.

To investigate the continental shelf efficiently, a model was developed to identify high potential areas for archaeological sites within southeast Alaska. Paleoenvironmental reconstruction allowed for the creation of 500 calendar year time-slices of southeast Alaska from 11,000 cal BP to 16,000 cal BP. The variables (slope, aspect, coastal sinuosity, distance from paleo-stream, distance from paleo-lakes, distance from paleo-coastlines, and distance from known archaeological sites) are included in the predictive model. This model has been used to delineate survey areas for underwater archaeological surveys during two-field season within Shakan Bay, Alaska using marine geophysical survey (side scan sonar, multibeam sonar, and sub-bottom profiling) and minimal sub-surface testing using a van veen grab sampler.

Four models were statistically tested: a 10-meter resolution model of southwestern southeast Alaska at modern sea level; a 10-meter resolution of southwestern southeast Alaska combining the models created from 10,500 to 16,000 cal BP; a two-meter resolution model of the Shakan Bay area at 11,000 cal BP; and, a two-meter resolution model of the Shakan Bay area combining the models created from 11,000 to 16,000 cal BP. These models were statistically tested using archaeological sites that have been identified/recorded since the creation of the models. The models indicate high predictive utility and it is expected that further field testing will identify archaeological sites on the continental shelf in Shakan Bay.

1

Introduction

1.1. Introduction

How people migrated to the Americas is still a complicated issue. Much of the current debate revolves around the lack of evidence along the Northwest Coast of the American continent. Though the 'Clovis first' model has officially been put to rest (Erlandson 2013; Ives et al. 2013; Mulligan and Kitchen 2013; D. G. Anderson, Bissett, and Yerka 2013), others are still sceptical of the coastal migration theory (Potter et al. 2017, 2018). Where are the sites? Underwater; now we just have to prove that. The Hakai Research Institute in partnership with the University of Victoria has focused on finding sites in areas where the sea level change would have been minimal, a hinge point. They have been able to locate two sites that have sensationalized this research in the last few years with dates older than 13,000 cal BP (McLaren et al. 2018; Muckle and Gauvreau 2017; McLaren et al. 2015; Letham et al. 2016). In Oregon, on the southern Northwest Coast (NWC) researchers unearth coprolites with human DNA that date to 14,000 to 14,270 cal BP[1] (12,300 ^{14}C years) at Paisley Caves (Gilbert et al. 2008). However, these recently uncovered archaeological sites are still not early enough for people to travel from Siberia to Chile at Monte Verde by 13,900 to 14,220 cal BP (Dillehay et al. 2008). We know that early people on the NWC are maritime adapted (Lindo et al. 2017; Erlandson et al. 2015; Braje et al. 2017; Sandweiss et al. 1998; Balter 2008; Q. Mackie, Fedje, and McLaren 2018). Therefore, the earliest sites should be along the coastline as it was when people migrated to the Americas. That coastline can be as much as -165 meters below modern sea levels in southeast Alaska.

1.2. Project Background

The Gateways to the Americas I and II projects, of which this research was a part of, aims to develop reliable methods to identify preserved archaeological sites on the continental shelf, supporting the coastal migration theory. Gateway to the Americas I was funded by the United States National Science Foundation (NSF)-Office of Polar Programs (OPP) (project number 0703980) to investigate the feasibility of testing the continental shelf in southeast Alaska for archaeological sites including one year of field research (Dixon and Monteleone 2011). Fieldwork was conducted in June 2010. Gateway to the Americas II was a multi-year project also funded by the US NSF-OPP (project number 1108367) including three years of fieldwork from 2012 through 2014 (Monteleone and Dixon 2018). The area covered by the Gateway to the Americas project encompassed the southern part of the Alexandra Archipelago from Prince of Wales Island west to the Pacific Ocean. This area is referred to as the 'study region' (figure 1-1).

The research presented here was originally the author's dissertation, a small part of the Gateway to the Americas projects. Here the focus is on Shakan Bay. Shakan Bay is one of nine identified sub-regions within the study region. Shakan Bay is identified as the 'study area' (figure 1-1). This research has continued to evolve since the completion of the author's PhD in 2013 (Monteleone 2013). What is presented, is an update including new analyses are more refined scales and using new and better technology.

The primary objectives of this research are to develop and test an archaeological predictive model designed to identify areas of high archaeological potential on the continental shelf of southeast Alaska. Sea level history and glacial geology suggest that the archaeological record prior to 10,600 cal BP (9,400 ^{14}C years) is submerged on the continental shelf in this region. To locate and test for sites older than 10,600 cal BP, it is essential to extend the archaeological record to the continental shelf in areas that were either unglaciated refugia or areas that were deglaciated during the interval between 16,000 and 10,500 cal BP. This research facilitates the exploration and the interpretation of the origins and character of early maritime adaptations along the NWC.

Archaeological sites document continuous occupation of southeast Alaska following sea level rise above modern levels around 10,600 cal BP. In southeast Alaska, a parallel study has been conducted by Carlson and Baichtal (2015; Baichtal and Carlson 2010) to investigate sites on raised marine terraces. This project has been extremely successful locating numerous sites between 6,000 and 10,600 cal BP. These early sites, which date between 13,300 and 10,000 cal BP, demonstrate that the region was occupied at times of lower sea level (Dixon 1999; R. J. Carlson and Baichtal 2015; R. J. Carlson 2007; Moss 2011, 2004; Wang et al. 2017). Coastal sites on the British Columbia, Canada (BC) mainland north of Vancouver Island have been identified dating to greater than 13,300 cal BP (Muckle and Gauvreau 2017). The expanding archaeological literature for pre-10,000 cal BP sites along the northern NWC supports the possibility of submerged coastal sites located on the continental shelf where maritime subsistence resources were abundant.

---

[1] cal BP = calibrated using Calib 6.0 (http://calib.qub.ac.uk/calib/calib.html) IntCal 09 (Reimer et al. 2009)

Uncovering Submerged Landscapes

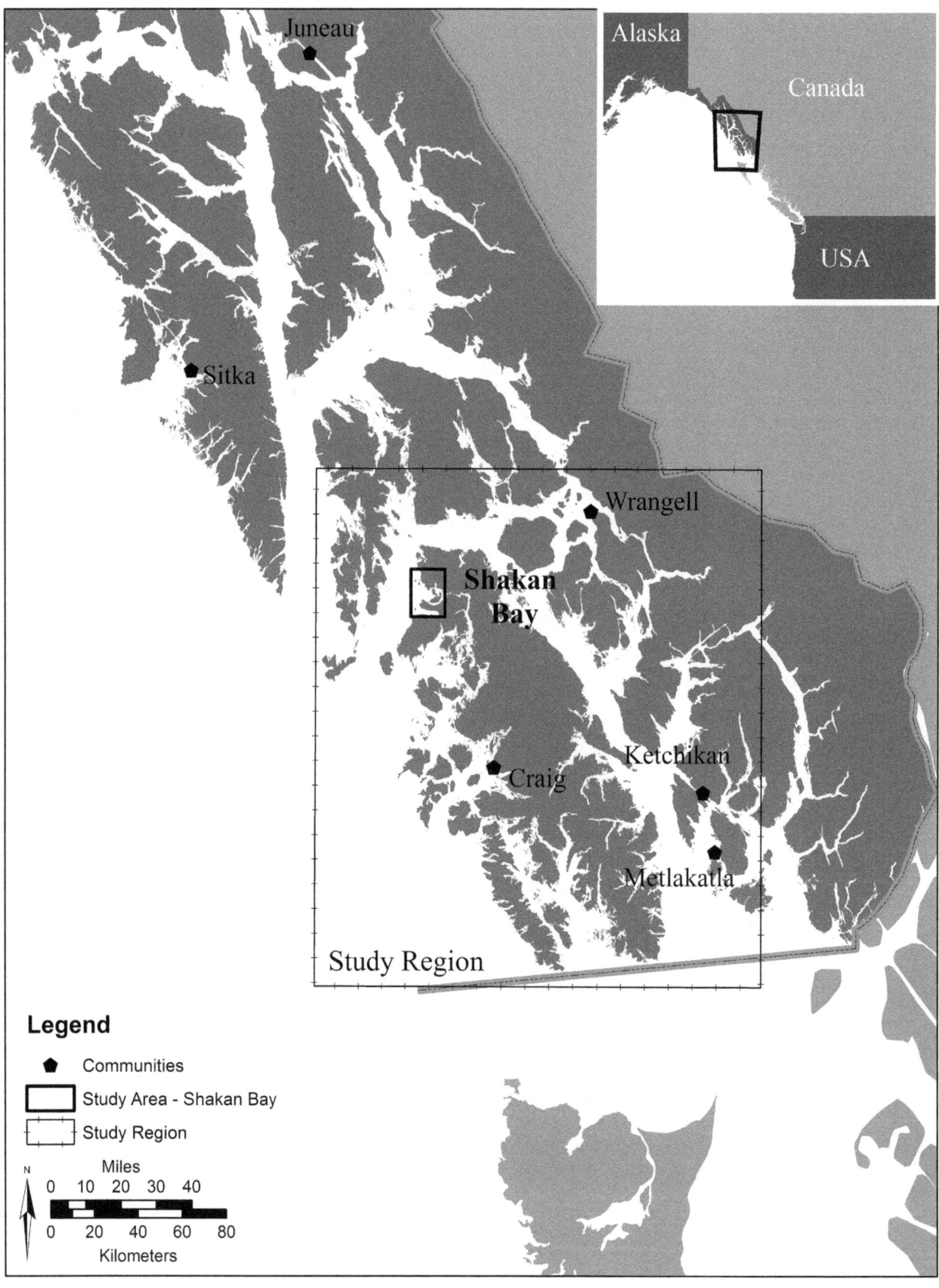

Figure 1-1. Map of study region and area.

The hypothesis this research is testing is that the archaeological record of Southeast Alaska extends to areas of the continental shelf that were submerged by post-Pleistocene sea level rise from 16,000 to 10,500 cal BP. People chose to live along the coast to exploit marine resources; they would have moved progressively landwards as sea level rose. An important goal for this project is to limit the area of the continental shelf for marine archaeological survey to increase the likelihood of locating a submerged archaeological site. An archaeological land-use or high potential model was developed using the geographic information system (GIS) function of weighted overlay in ESRI's ArcGIS. The variables incorporated are slope, aspect, sinuosity, and distance from streams, lakes, tributary junctions, the coast, and known archaeological sites. The model was produced at 500 cal BP intervals. For the study region, the model was 10-meter resolution and spanned 10,500 to 16,000 cal BP. For the study area, a two-meter resolution model was developed from 11,000 to 16,000 cal BP.

This is the first project to explore the continental shelf in southeast Alaska and one of a few in North America (see chapter two: underwater archaeology). It was an exploratory project with a low probability of locating archaeological sites in the limited time available. This is partially due to:

1. The need to refine methods for survey and excavation,
2. The large expanse of the continental shelf within the study region,
3. The scarcity of early hunter-gather archaeological sites globally,
4. The probable small size of possible sites, and
5. The expanses of time since these sites were deposited into the archaeological record.

Though the underwater environment can provide excellent preservation for organic and lithic artefacts, taphonomic processes will limit the length of time an archaeological site will remain in an identifiable state on the seafloor. The predictive model is intended to focus the survey areas to high potential locations. These high potential areas can be generalised as 'flat areas near freshwater and the coast'; or, areas where past people would have used the landscape for resource extraction. Potential problems include the scarcity of available input information (more detailed bathymetry and topography would produce a more precise model), changes in the environment over the last 16,000 years (including storms and tectonic events which may have destroyed or moved possible archaeological sites), difficulties during marine geophysical surveys and sampling, and the vast area to be surveyed and sampled. The study region is over 45,000 km²; the study area is 1100 km².

A quick note about tectonic processes: though southeast Alaska has two major faults, the Queen Charlotte-Fairweather fault and the Chatham Strait fault, tectonic processes are not incorporated into the model. Tectonic events can cause several meters change in elevation over local areas; these are not predictable. However, as these changes in elevation would be reflected in the sea level, tectonic changes are incorporated as part of the overall land-sea level reconstruction.

1.3. Study Region and Study Area

The study region is the southwestern Alexander Archipelago in southeast Alaska (figure 1-1) in the northern NWC of North America. Recent research indicates southeast Alaska and western British Columbia were largely glaciated beginning around 21,000 to 17,000 cal BP (18,000 to 14,000 ^{14}C years) albeit with refugia (unglaciated areas capable of supporting humans) existing along the coast (Carrara et al. 2003; Carrara, Ager, and Baichtal 2007; Clague, Mathewes, and Ager 2004; Shugar et al. 2014). By 16,000 cal BP (13,500 ^{14}C years), much of the region was deglaciated and ecologically viable for human habitation, although a few valley glaciers from the Coast Mountain Range still extended to the coast (Carrara, Ager, and Baichtal 2007; Clague, Mathewes, and Ager 2004, 94; Mathews 1979; Shafer et al. 2010). Sea level reconstructions for Haida Gwaii, formerly named the Queen Charlotte Islands (Q. Mackie, Fedje, and McLaren 2018; Fedje et al. 2018) and southeast Alaska (Dixon and Monteleone 2014; R. J. Carlson and Baichtal 2015; Monteleone 2013) indicate sea level has risen approximately 165 m since the last glacial maximum, briefly rising above modern sea level in many areas around 10,600 cal BP. Although the character of the NWC continental shelf within the study area varies, it extended from three to 50 km (one to 30 mi) seaward from the mainland coast prior to 10,600 cal BP.

Within the study region, Shakan Bay was identified as a high potential location for archaeological sites. The larger Gateway to the Americas projects has identified several study areas for survey. These include Keku Strait, the Gulf of Esquibel, and Suemez Island. Keku Strait was selected based on a report from a local fisherman of an artefact recovered from the seafloor. Suemez Island was selected based on its proximity to a known obsidian source and palaeoenvironmental reconstructions that indicate that the southern cove may have been part of larger ice-free refugium during the LGM (Carrara et al. 2003; Carrara, Ager, and Baichtal 2007; Moss and Erlandson 2001). Shakan Bay was identified because it would have been an intertidal estuary from at least 12,000 to 13,000 cal BP and a series of connected lakes from 13,000 cal BP. If the area were not covered in glaciers during the LGM, the connected lakes would have remained as the seafloor drops quickly and steeply into Sumner Strait. These past environments would have been highly productive for past peoples similar to the Ertebølle culture in northern Europe. In addition, Shakan Bay is sheltered by Baranof Island to the west and does not receive the full fetch of the Pacific Ocean, but Sumner Strait, to the west, is deep and would have support diverse marine resources. Because there are elevation changes within the bay that allows for hills, valley, and depressions to form, Shakan Bay would have been an excellent place to live at all sea levels.

1.4. The GIS Model

High potential areas are defined based on the synthesis and interpretation of archaeologically and ethnographically documented land-use patterns applied to reconstructions of the submerged landscape. The model incorporates both inductive, utilising known archaeological site data, and deductive, utilising anthropological theory and the ethnographic record, types of modelling (Verhagen and Whitley 2012). The scale of measurement for this analysis are interval or ratio, and both the analytic (archaeological) and the systemic (dynamic living system) contexts were analysed to develop the model (Kohler 1988, 35–37; Schiffer 1972).

The analytic units are the ranked valued for each of the model variables. The synthetic units are the larger groupings of the variables. These are slope, aspect, water variables, and sites. Slope and aspect defined as specific location properties, based on rank. Water variables include distance from streams, lakes, the coast, and tributary junctions. This variable type not only provides access to freshwater but also is the main location for dietary resources and for transportation. Sinuosity is incorporated into the synthetic unit because it is combined with distance from the coast. Sinuosity reflects the amount of coastline available from a location. Increased sinuosity often indicates more resources available (A. P. Mackie and Sumpter 2005). The last synthetic unit sites and incorporates distance from known archaeological sites. This increase the potential for archaeological sites near known sites Binford's (1980) logistical collectors, whereby people move to nearby camps, stations, and caches. People exploit resources near each of these site types. Camps, stations, and caches then become sites and new sites will be 'near' other sites or camps. This is also referred to as clustering.

Summary statistical are a means to estimate appropriate weights for theoretically derived variables (Kvamme 2006, 12). Statistics were generated from known archaeological site locations. These statistics were then qualitatively compared to ethnographic accounts and general anthropological theory. For example, a theory such as 'people live near freshwater' was compared to the median distance from streams, lakes, and tributary junctions. Because the median values for lakes and streams were closer to 500 m than 100m, the highest rank was assigned to the 500 m buffer for these variables. This means that people still live near water, but more often, they were a short distance from these sources of freshwater. Based on these qualitative assessments, the variables in the model were ranked and were weighted out of 100 per cent. This model uses inductive modelling techniques to derive deductive variables (Kvamme 2006, 12).

1.5. Peopling of the Americas

This model has the potential to test and provide evidence for the coastal migration theory. This theory postulates that the NWC of the Americas was a possible route that the first Americas took when migrating to the New World (Dixon 1993, 1999, 2011a, 2013a, Fladmark 1975, 1979; Gruhn 1994; Heusser 1960; Braje et al. 2017; Erlandson et al. 2015). It is hypothesised that these migrants would have used watercraft to travel along the coastal areas of the north Pacific Rim that are now submerged by post-Pleistocene sea level rise. This means the continental shelf of southeast Alaska may contain archaeological sites from the earliest settlers in the Americas.

The 'ice-free corridor' proposed between the continental glaciers was not viable for human occupation until 13,000 cal BP (11,000 ^{14}C years). This has been corroborated by several independent researchers from different disciplines including archaeology, geology, and palaeontology (Arnold 2006; Dixon 2011a; Mandryk et al. 2001; Wilson 1996; Potter et al. 2018, 2017; Freeman 2016). As Freeman (2016) points out, this late date does not mean that the deglaciated corridor should not be investigated as a migration route. Evidence for the Atlantic coastal migration is still lacking genetic and linguistic support (Stanford and Bradley 2012; Collins et al. 2013). This leaves the coastal migration theory as a leading theory in explaining the first colonisation of the Americas. This is supported by Wheat's (2012) survey of scholars researching the peopling of the Americas where 86 per cent favoured the coastal migration theory contributing to the initial peopling of the Americas. All major areas along the Pacific Coast of the Americas were occupied by at least 13,000 and 11,500 cal BP. These occupations are contemporary with Folsom and Clovis sites in the continental interior (Erlandson, Moss, and Des Lauriers 2008; Dixon 2011a).

1.6. Northwest Coast Maritime Adaptations

Along the NWC, the timing of intensification of maritime adaptation is an ongoing discussion, specifically if the early inhabitants were full maritime adapted and exploiting a maritime culture, or if they developed these technologies after moving to the coast. A maritime culture is adapted to the sea and acquires the majority of its dietary and economic needs from the sea (McCartney 1974, 158–59). The non-maritime adapted theory assumes that people moved into the NWC from riverine environments and then later developed technology for salmon harvesting and deep-sea fishing. This maritime culture was then only fully developed by approximately 5800 calendar years ago (Borden 1950, 1951; Fladmark 1979; Kroeber 1939; Potter et al. 2017; Moreno-Mayar et al. 2018).

The fully adapted maritime view assumes the cultures that migrated into the NWC were already utilising the sea (Borden 1975; A. Cannon 1998; Drucker 1955; Q. Mackie, Fedje, and McLaren 2018; Fedje et al. 2001). This theory is synergistic with the coastal migration theory. Early evidence for maritime adapted cultures along the Pacific Coast can be found in the Channel Islands by 13,000 cal BP (Moss 2011; Rick et al. 2005; Erlandson et al. 2011). Similar to some of the Channel Island evidence, there is overwhelming evidence on the Northern NWC of travel by boat. On Haida Gwaii, there are several sites dating to 12-13,000 cal BP when Haida Gwaii would have been an island chain. Obsidian recovered from OYKC dating to older than 10,000 cal BP is sourced to Suemez Island, a distance that could not have been travelled by land 10,000 cal years ago (Dixon 1999, 2013a; Dixon et al. 1997; Lee 2007). Early maritime adaptation is supported by recent evidence including site location models on Haida Gwaii (A. P. Mackie and Sumpter 2005), human remains from On Your Knees Cave with an isotopic signature of a marine carnivore (Lindo et al. 2017; Barrie and Conway 1999), and consistency in salmon harvesting since at least 7000 cal BP at Namu (A. Cannon and Yang 2011).

The kelp highway is a ring around the northern Pacific, from Japan through California/Mexico, where similar resources could be exploited (Erlandson et al. 2015, 2007; Erlandson 2013; Braje et al. 2017). Coastal migration theory along the kelp highway would have provided a linear migration corridor with few barriers to maritime peoples. This path would have had abundant resources in the nearshore, estuarine, riverine, and terrestrial ecosystems that would have allowed populations to rapid grow and provide demographic pressures to fuelled exploration and colonisation (Erlandson et al. 2015, 408). This route would have been open through to northern Vancouver Island by 16,000 cal BP, where a 400 km stretch of coast was still glaciated (Dixon 2013b). Based on the available evidence, the coastal migration theory is the reasonable choice; but Potter and colleagues (2017, 2018) still need more evidence that early people along the NWC were maritime adapted. This research aims to gather more data about the early NWC and maritime adaptations.

1.7. How Does This Relate to Other Projects?

Chapter two reviews the history and other projects where predictive models are utilised and where underwater archaeology is employed. This research is rare that it utilises both a formalized predictive model and underwater archaeological methods to locate submerged archaeological sites. Other underwater archaeological predictive modelling projects that have developed interval/ratio scale measurements using inductive and/or deductive methods include Barbra and Roberts in 1979 for the northeast United States and Muche's 1981 model for Santa Monica Bay, California. Many underwater archaeology projects include aspects of predictive modelling including mapping specific environmental/geologic variables and sea level changes. Numerous projects reconstruct paleoenvironments to identify locations of high probability for archaeological sites. However, these projects are infrequently or discretely using the predictive modelling literature as a source to narrow the search for sites on the seafloor, i.e. they are using inductive or deductive reasoning but not developing interval or ratio scale measurements.

Archaeological predictive modelling is an extensive subfield within archaeology. The methods and theory behind archaeological predictive modelling have expanded extensively with the ready availability of GIS. Options and possibilities of open source GIS software make spatial analysis available to everyone, not just funded projects. With the introduction of ArcPro, an ESRI product for analysis and mapping, processing times are getting even shorter allowing for analyses that are more complex. ArcPro is the first ERSI tool to utilise 64-bit technology and multiple computer cores. This project is somewhat unique in the resolution and scale of analysis. Most analyses at two-, five- or 10-meter resolution are not at the scale of 45,000 km^2. In an earlier iteration of the model, the 10-meter resolution was actually too large for several analyses (Monteleone 2013). Some steps in this analysis took weeks to complete because of the computational limits during the first production prior to ArcPro.

1.8. Structure of This Book

Chapter two presents the theoretical background underlying this research. It overviews landscape theory as high-level theory, archaeological land-use or high potential modelling as middle-range theory, and GIS (geographic information systems) and underwater archaeology as low-level theory. Chapter three reviews the geographic and geologic background for southeast Alaska, the Alexander Archipelago, and Shakan Bay. Chapter four discusses the archaeological and ethnographic literature for southeast Alaska and the northern Northwest Coast (NWC) in terms of research questions. The region's three Native American ethnographic groups each are reviewed in relation to traditional land-use, seasonal round, social structure, and economic orientations. Chapters five, six, and seven, present the high potential model, and its statistical and geophysical testing. The results of two field seasons of marine geophysical survey and subsurface testing are reported in chapter seven. Chapter eight summarises archaeological sites that are older than 9,000 cal BP on the northern NWC, their relevance to this research, and how these sites are located in relation to the land-use model. Finally, chapter nine discusses the geologic history of Shakan Bay based on the multibeam and ROV results, the implications of model resolution on the results, the survey results and future directions.

2

Theoretical Background

2.1. Introduction

The theoretical framework has been divided into high, middle, and low theory based on Schiffer (1988). 'Theory' is used generally in the sense of explanations, models, or hypotheses. Essentially this chapter is the theory for each of the four major background areas for this research: landscape, modelling, GIS, and underwater archaeology.

Landscape theory is the overarching or high-level theoretical framework for this research. It provides a theoretical framework whereby the research focus is appropriate for an area that is larger than an archaeological site. It facilitates analysis at multiple scales, incorporating regional geomorphology and actualistic studies (e.g. site formation processes and ethnoarchaeology) to answer questions regarding land-use, site distribution patterns, and other spatially related questions (Anschuetz, Wilshusen, and Scheick 2001; Bender 2002; Casey 2008; Kantner 2008; Rossignol and Wandsnider 1992a). High potential archaeological site location or land-use models are a form of middle range theory (Verhagen and Whitley 2012). GIS and underwater archaeology provide methodological or low-level theoretical frameworks. Though the low-level theories are uncomplicated explanations of methods background, they still form a theoretical background.

2.2. Landscape Theory

Modern archaeological landscape theory has its roots in Julian Stewart's concepts of cultural ecology. It is a modern embodiment of human and historical ecology and relates to evolutionary ecological models. *Ecology* is the study of the relationship between organisms and their environment (Dincauze 2000, 3). The *environment* is 'all physical and biological elements and relationships that impinge upon a living being' (Dincauze 2000, 3).

Landscape archaeology is protean. It provides a conceptual framework to address all contexts of past human behaviour and goes beyond an 'environmental' approach. It is focused on things that locate humans spatially (David and Thomas 2008). Landscape archaeology is about 'place' as a basic unit of lived experience (Casey 2008, 44). Descartes (1991, 45–46) distinguished 'place' as a position versus 'space' as a volume. Following Descartes, 'landscape' is inconceivable without place. It is a set of discrete 'places' (Casey 2008, 49).

> Landscape archaeology is an archaeology of how people visualised the world and how they engaged with one another across space, how they chose to manipulate their surroundings or how they were subliminally affected to do things by way of their locational circumstances. It concerns the intentional and the unintentional, the physical and the spiritual, human agency and the subliminal. (David and Thomas 2008, 38)

Landscape theory includes 'taskscapes' (Ingold 1993) and 'seascapes' (Bjerck 2009; Chapman and Chapman 2005; Van de Noort 2003). *Landscape* is a cultural construction of modern European society. It is a western philosophical framework being applied to a North American traditional knowledge base (Bender 2002, s105; Knapp and Ashmore 1999, 6).

There are three tenets central for archaeological landscape theory. First, there is a dedicated effort to examine the physical environment using various natural science techniques focusing on scientific questions (Feinman 1999, 684; Rossignol and Wandsnider 1992a). Secondly, archaeological landscape theory recognises distinct cultural perceptions and past human actions shape human-environment interactions (Anschuetz, Wilshusen, and Scheick 2001; Bender 2002; Feinman 1999, 684). Finally, landscapes are products or creations of a dynamic interaction with human behaviour and the natural environment, to a certain degree (Bender 2002; Feinman 1999; G. G. A. G. Johnson 1977; Kantner 2008; C. R. Fischer and Thurston 1999). 'Landscapes in this conceptualisation are the intersection of a particular group's history with the places that define its spatial extent' (Anschuetz, Wilshusen, and Scheick 2001, 186). Hence, a landscape is a specific location during a specific time to a specific group of people.

2.2.1. Historical Development of Landscape Theory

Early (pre-Boasian) geographic and anthropological works focused on the effects the environment had on the development of people, or sometimes the reverse. Environmental determinism created simplistic correlations between environmental and cultural features at large-scales (Kohler 1988, 25). Boas recognised the importance of environment and correlated land-use practices with environmental features by reducing the scale of analysis to assess environmental features and aspects of human behaviour as it pertained to specific groups of people (Boas 1935; Kohler 1988, 25). In developing the culture area concept, Kroeber and Wissler minimise the effects of causality from the environment and cultural relationships (Kohler 1988, 25–26; Wissler 1917; Kroeber 1923). In the 1920s, Carl Sauer, a geographer, recognised that landscape is the repository of a culture's impact on its environment. A landscape is a tangible record of human adaptation to the physical environment (Anschuetz, Wilshusen, and

Scheick 2001, 164; Knapp and Ashmore 1999, 3). Many archaeologists viewed past people as actors whom either: [1] adapted to their environment, [2] transformed their environment, or [3] they became extinct. Today, many archaeologists assume, intentionally or unintentionally, that the modern landscape was more or less representative of past landscapes without any significant changes (Patterson 2008, 78). This is not an assumption of this study.

Julian Steward (Kroeber's student) was interested in causal explanations, not correlations, using the culture-area concept. He emphasised the effect of local aspects of the environment on particular facets of culture, not large-scale correlations of regional environments with culture types. The 'culture core' concept, defined as 'the constellation of features which are most closely related to subsistence activities and economic arrangements' (Winthrop 1991, 47), was used as a pathway through which environment might influence cultures (Kohler 1988, 26). Steward was then able to extend the culture-core concept to specific site location models. He generalised: some groups in arid portions of western North America were highly responsive to a limited number of mappable environmental determinants (Kohler 1988, 27–28). Steward's views on cultural ecology suggest that settlement behaviour is part of the 'culture core'. Because there is no clear objective procedure for determining what constitutes the core, Steward assumes the environment influences culture, but culture does not influence the environment (Kohler 1988, 28–29). Trigger (1971) refers to this as 'deterministic ecology'. In response to the problems with Steward's methods, many archaeologists moved to an ecosystem approach influenced by evolutionary ecology (Flannery 1968; Winterhalder 1981).

Steward's student, Gordon Willey, started settlement pattern archaeology as a new field of inquiry beginning with the survey of the Virú Valley in Peru. 'The term "settlement pattern" is defined here as the way in which man disposed himself over the landscape in which he lived' (Willey 1953, 1). Willey associated settlement patterns with environmental, technological, and demographic change (Kohler 1988, 30; Patterson 2008, 78). After Willey's study, archaeological survey methods became increasingly rigorous, applying regional and sub-regional scales of analyses (Anschuetz, Wilshusen, and Scheick 2001, 169).

In the 1960s and 70s, there was a divergence in landscape approaches. The first was a positive science emphasising spatial quantitative approaches. These include network, hierarchical, and surface models, such as Hodder's (1977) macro-scale analysis. The second approach focused on humanistic philosophies and human values, beliefs and perspectives, including idealism, feminism, and phenomenology (Anschuetz, Wilshusen, and Scheick 2001, 164–65).

In the 1970s, the Southwestern Anthropological Research Group (SARG) introduced more rigorous testing to determine the degree of relationship between site location and environmental variables. They incorporated expected site distributions for comparison with observed site distribution using formal statistical inferential techniques including analysis of variance (ANOVA). The final theme in their report focuses on the difference between archaeological site locations and the distribution and movement of people through space in the systemic behavioural context. '[P]rehistoric peoples most likely did not locate 'sites' anywhere. However, they did establish, occupy, and abandon behaviourally significant spaces, such as activity areas, camps, and settlements' (Sullivan and Schiffer 1978, 169).

Citations began for landscape archaeology as a specific term in academic work in the mid- to late-1980s. In the 1980s, landscape changed from simply being a unit of analysis, which is larger than a site, to a high-level theoretical approach (David and Thomas 2008).

2.2.2. Modern Landscape Theory

Landscape theory is evolving and dynamic. However, there is and has been a difference between European and North American views of landscapes (Lock and Harris 2006). This research incorporates both views of landscapes. Landscape archaeology incorporates land-use patterns, subsistence-settlement systems, ecological habitats, both terrestrial and celestial landscapes, materialisation of worldview, built or modified (culturally constructed) environments, and stages for performance (Patterson 2008, 77). These theories are based in phenomenology, ecology, subsistence theory, and practice, though many other interpretive schemas exist. Landscape theory is able to accommodate and integrate different theoretical perspectives, even those that create competing reconstructions of the past (Anschuetz, Wilshusen, and Scheick 2001, 159, 176).

Landscape archaeology has expanded beyond Steward's descriptive studies to address observed patterns generated across space and time (Anschuetz, Wilshusen, and Scheick 2001, 170; Kohler 1988; Kohler and Parker 1986, 399; Patterson 2008, 78). Predictive modelling seeks to build from the relatively static spatial distribution of material culture and anthropogenic alterations visible in the modern landscape to an understanding of cultural dynamics and environmental process (Bevan and Conolly 2005, 218). It incorporates quantitative methods. It explores the correlation between landscape and social or environmental variables and investigates neighbourhood dependence, which is the physical relationship between new and existing sites or households. These quantitative methods can be referred to as predictive modelling (Bevan and Conolly 2005).

Phenomenology interprets landscapes from the point of view of the subject. A landscape has to be experienced to be understood. The phenomenological objective is 'to provide a rich or "thick" description allowing others to comprehend these landscapes in their nuanced diversity and complexity' (Tilley 2008, 271). The main research tool for phenomenology is the physical body or the medium of the researcher's senses. A landscape can have agency. It can act on a person by affecting thoughts and interpretations. Landscapes can be material. They are real and physical, not only a cognitive or imagined reality. As discussed earlier, landscapes exist in space-time. They are always changing and are never the same twice. Phenomenology seeks to ascertain how others experience a landscape through learning how others see and experience the landscape (Tilley 1994, 2004, 2008). This research recognises the importance of phenomenological studies, but acknowledges that the submerged continental shelf, physically, only can be experienced through digital reconstructions. Thus, this research is investigating a virtual cultural landscape. There are several arguments against phenomenology. The most notable argument is the 'highly questionable results' of fieldwork (A. Fleming 2006, 267). Phenomenology provides a broader mentality for this research. The landscapes are reconstructed and visualised to attempt to experience the landscapes using phenomenological methods; how would people have viewed their landscape in the past.

Evolutionary ecology and human behavioural ecology (HBE also referred to as behavioural ecology) provide a theoretical platform upon which landscape archaeology has been applied in this research. Evolutionary ecology is 'the application of natural selection theory to the study of adaptive design in behaviour, morphology, and life history' (M. D. Cannon and Broughton 2010, 1). HBE is a sub-area of evolutionary ecology that is most directly relatable to archaeology. 'HBE focuses on functional variability by applying natural selection theory to the study of human behaviour in socioecological context' (D. Bird and Codding 2008, 396). It draws on formal optimisation and game theory models to develop hypotheses for fitness related trade-offs actors may face in variable environments and circumstances. Evolutionary ecological models use assumptions relevant to the evolution or expression of a phenotypic variant, whereas HBE focuses on behavioural components of the human phenotype. The evolution and expression of a behaviour are affected by the natural and social features of the environment (M. D. Cannon and Broughton 2010, 2; D. W. Bird and O'Connell 2006, 143; Shennan 2002). People made decisions based on underlying logic to sustain the 'group' or social system. At the scale of the archaeological record, these unconscious or conscious decisions to maintain homeostasis (or conscious decision to disrupt homeostasis) are what are discernible. The units of analysis are at the levels of group, social system, or society, even though the aim is investigating how choices affect individuals' survival and reproductive success (Bettinger, Garvey, and Tushingham 2015; D. Bird and Codding 2008; G. G. A. G. Johnson 1977; Kantner 2008). This means that HBE's goal of determining how ecological and social dynamics affect behavioural decisions within and between groups is expressed through the archaeological site locations.

Landscapes are the context in which decisions or behavioural choices are made. These choices affect an individuals' survival and reproductive successes (D. Bird and Codding 2008; G. G. A. G. Johnson 1977; Kantner 2008). 'They [choices] form the ecology of human life – the social history, individual development, and local environmental circumstances that constrain our behaviour' (D. Bird and Codding 2008, 396). HBE attempts to explain complex cultural patterns based on individual interactions and decisions. The two basic assumptions are: [1] resources are finite and individuals make trade-offs because of resource finality; and [2] natural selection determines the capacity to evaluate the cost and benefits of these trade-offs (D. Bird and Codding 2008, 397; Shennan 2002). Research involves two tasks: reconstructing behaviours from the archaeological record and explaining the reconstruction (D. W. Bird and O'Connell 2006, 144). This research evaluates a regional archaeological predictive model in terms of behaviour choices utilising HBE.

Behavioural choices or 'cognitive variables' are the difference between an 'ecologically deterministic' and a humanistic approach (van Leusen et al. 2005, 30). These decision-making processes make hunter-gatherer subsistence economies uniquely suited to landscape and GIS studies (Boaz and Uleberg 2000). For this research, the behavioural choice modelled is subsistence by marine resources. This means that there is a greater preference for coastlines with access to coastal resources, and stream, lakes, and tributary junctions where other marine resources, such as anadromous fish, are accessible.

Subsistence or economic landscapes are amenable to investigating the distribution of resources and interpreting economic strategies. Landscape-scale ethnoarchaeological research investigates the organisation of subsistence strategies at the regional level, how these strategies shift through time, and how organisation influences the structure of the archaeological record (Stern 2008; Rossignol and Wandsnider 1992b, 7; Wandsnider 1992a). Ethnographic research was vital to developing a model appropriate for the continental shelf of southeast Alaska to incorporate cultural decision-making. This model maps changes in coastlines, streams, lakes, and tributaries but implies that these locations are where the resources were located. This means that by changing these variables, the location of past resources is being mapped. Whitley and Burns (2008) developed a high potential model for South Carolina incorporating resources and the distance to these resources. It utilises a combination of land-use and subsistence resource patterns to develop the model.

Though in the past some archaeologists have viewed landscapes as stagnant, contemporary landscape theory incorporates the environmental and cultural changes that have occurred through time (Patterson 2008, 78; Rossignol and Wandsnider 1992a). This requires environmental reconstruction for each time-space investigation (Tilley 2008). Modern landscapes are the culmination of past landscapes or 'time materialised' (Bender 2002). Many researchers do not reconstruct the environment and this is a serious flaw in many landscape studies. Ideally, reconstructions will incorporate regional geomorphology and actualistic studies. Actualistic studies examine present-day, dynamic systems, which can be related to the deposition of archaeological remains. These come in three forms: [1] inquiry into formation of the fossil assemblages (taphonomy), [2] investigation of the effects of interactive human and natural processes on archaeological assemblages, and [3] examination of the impact of hunter-gatherer and agropastoral mobility strategies of the landscape (Rossignol and Wandsnider 1992a, 6). Another actualistic approach is ethnoarchaeology that explores anthropogenic impacts on archaeology within systemic regional patterns of human behaviour in space and time (Rossignol and Wandsnider 1992a, 7). For this research, many early archaeological sites have been located in caves and in intertidal areas. This means that older sites are not often preserving in open environment or surviving. The underwater environment can provide an anaerobic environment to preserve archaeological site remains. The taphonomic processes at work underwater are complex. The underwater environment can be more destructive than most open-air environment, but by simply covering a site by a few centimetres of sediment, a site can be preserved almost indefinitely, such as Bouldner Cliffs in England and La Mondrée in France (Momber 2011, 2000; Cliquet et al. 2011). Chapter three specifically deals with the underwater environment and the interactions that the waves, tide, and other processes have on submerged environments through time. Finally, because of the few sites that have been located prior to 10,000 cal BP in the region (chapter eight), it is unknown what effects the people of the NWC had on the environment that is now submerged. Rock aliments for fish weirs, depressions or house pits, caches, and shell middens are all expected on the seafloor. Unfortunately, the amount they are buried will affect both their preservation (positively) and our ability to identify them (negatively). The changes expected from actualistic studies for the underwater environment along the NWC are inconsistent and difficult to determine. This research anticipates that some areas will still contain submerged archaeological sites that are identifiable.

2.2.3. Non-sites

In the early 1980s, Dunnell and Dancey (1983) questioned the usefulness of 'site' as a unit of analysis. This was part of the growth of landscape archaeological theory (Patterson 2008, 78–79). People did not choose to locate sites. They interacted with the environment and landscape. Archaeologists identify archaeological sites and define their boundaries. Sites are areas where artefacts are found (Willey and Phillips 1958, 18). Dunnell and Dancey (1983, 272) suggest, 'the archaeological record is most usefully conceived as a more or less continuous distribution of artefacts over the land surface with highly variable density characteristics'. The density of artefacts characterises land-use and reflects its frequency. Willey and Phillips (1958, 18), while defining the unit of 'site', recognise that a site is 'often impossible to fix'. They mean that the exact location of a site is difficult to determine.

Dunnell and Dancey's (1983) concept of non-site opens the entire landscape or region for analysis and survey. Clusters, or densities, of artefacts or human modification of the landscape, can be utilised to interpret the past. Even without the limitation of the concept of sites, data can be organised into sites as classificatory units. Ebert and Kohler (1988: 144) point out that looking for non-sites would be very expensive, and suggest, 'archaeological sites... are rare phenomena that only occur "about one per cent of the time".' From this perspective, sites are viewed as nodes in a continuous distribution of archaeological material. Therefore, one out of 100 random points distributed through a landscape, will actually be an archaeological site by chance. Despite their resistance to the concept, Ebert and Kohler (1988, 145) support a 'non-site' sampling approach. By incorporating the premises of non-site landscape analysis, the minimum operating units of archaeological analysis are artefacts and features. This research incorporates the non-site concept for survey. Sites and site boundaries will be difficult to determine on the seafloor; any distribution of artefacts would equate to a positive response during survey for this model. Maritime archaeology has focused on shipwrecks, which are sites; however, underwater archaeology has focused on individual finds such as faunal remains and a single utilised flake from Haida Gwaii (Fedje 2000; Fedje, Christensen, et al. 2005). Essentially, this model is trying to locate archaeological remains and features and is not specifically looking to locate archaeological sites. The model specifically focuses on land-use or areas where people could have collected the maritime resources they required for subsistence and economic purposes.

2.2.4. Seascapes

McNiven (2008, 150) defines seascapes as views from land to sea, views from sea to land, views along a coastline, and the effect on a landscape of the conjunction of sea and land. Seascapes are viewed as synergistic incorporating both landscape and maritime archaeology (Bjerck 2009; Van de Noort 2003, 405). Seascapes incorporate names, myths, legends, and property or political boundaries vital to the identity of maritime people. The sea can be anthropomorphised and thus imbued with agency and intentionality (McNiven 2008, 150). Seascapes also can have interesting phenomenological interpretations, but 'seascape' often is used interchangeably with landscape when referring to coastal areas (Bjerck 2009; Breen and Lane 2004; Chapman and Chapman 2005; Van de Noort 2003).

Ford (2011a, 4) defines seascapes as 'constructed of the factors that allow an individual to perceive his or her location out of the sight of land'. This incorporates aspects of astronomy and mental maps integral to navigation (Lewis 1972). Westerdahl (1992) defined the 'maritime cultural landscape' as the remnants of maritime culture on land and underwater. The maritime cultural landscape incorporates 'human utilisation (economy) of maritime space by boat: settlement, fishing, hunting, shipping and its attendant subcultures' (Westerdahl 1992, 5). It can be integrated in the seascape concept and landscape theory as it applies to settlement, land-use, or high potential modelling.

The seascape concept can be a much-generalised view. By combining seascape and phenomenology, this research aims to reconstruct the past (and now submerged) environment so that it can be viewed as a fluid landscape where people moved, often-utilising water transportation, viewing the land from the sea.

2.3. Archaeological Predictive Models

The goals of a land-use, settlement, or high potential model is to construct 'the distribution of related tolerances, that influenced the spatially contextual decision of humans in the past, which led to observed settlement patterns in the archaeological record' (Carleton, Conolly, and Ianonne 2012, 3371). A model can be regarded as a collection of irregular polygons, mapped onto a landscape, indicating locations that are 'favourable', 'likely', or 'probable' to contain an archaeological site (Kvamme 2006, 27) or where people choose to locate themselves in relation to resources of the type being modelled. Models to predict high potential, or probable, areas for archaeological sites have been viewed negatively by some academics because of the lack of theory and inappropriate application (Sebastian and Judge 1988). The applicability of archaeological site location models has been exaggerated in the past. This research uses the model solely to limit areas for survey. Identifying an area as high or low potential does not affect the development or preservation of that environment in this case. The problem with past predictive models is that they were viewed as a means to identify areas for development that were low potential and did not require further archaeological investigation.

A basic definition of *predictive modelling* is 'a technique that, at a minimum, tries to predict the location of archaeological sites or materials (non-site) in a region, based either on a sample of that region or on fundamental notions concerning human behaviour' (Kohler and Parker 1986, 400). Verhagen and Whitley (2012, 52–53) expand this definition to require the model to have 'a quantitative estimate of the probability of encountering archaeological remains outside the zones where they have already been discovered'.

A *model* is defined as 'hypotheses or sets of hypotheses which simplify complex observations whilst offering a largely accurate predictive framework structuring these observations' (Clark 1968, 32). First, this means that a model is a hypothesis that needs to be tested and must fit within the larger theoretical framework of landscape archaeology. Second, a model is an abstraction that can be generated inductively or deductively, but it always incorporates the subjectivity of the observer. Third, a model has predictive capacities. Hence, the term 'predictive model' is considered by Sebastian and Judge (1988, 1) as redundant. Nevertheless, it has predominated in the literature since the 1970s. Finally, a high potential model for landscape archaeology is a bridging function, or middle range theory, linking high-level theory to data and the archaeological record (Verhagen and Whitley 2012).

The niche concept from biology has been used as an analogy to archaeological high potential modelling (Kvamme 2006). Kvamme (2006, 14) defines *niche* as 'the total range of conditions in the environment under which a population lives and replaces itself'. This is a simplification of the original niche concept. Brown (1995, 30) states that a niche according to Hutchinson 'could be represented quantitatively in terms of the multidimensional combination of abiotic and biotic variables required for an individual to survive and reproduce'. Species exhibit distinctive patterns of abundance and distribution that reflects those species environmental requirements. For any species, 'ideal habitats' can be represented by measurements of specific variables, usually environmental variables. Human uses of space can be viewed in terms of a subset of environmental variation. Even culturally determined variability can be mapped using environmental variables, though this must be tested and supported in each case. Niche modelling provides a logical basis for this archaeological modelling (Kvamme 2006, 14) and is a recent application of archaeological high potential modelling (Kondo and Oguchi 2012; Kondo, Omori, and Verhagen 2012).

Predictive modelling in Europe is focused on three primary concepts: [1] sustainability of archaeological-historic resources; [2] method of non-destructive resource management, integrating the historical landscapes and buildings in environmental and spatial planning; and, [3] justifying heritage management decisions to future generations (van Leusen et al. 2005, 25; Verhagen 2008). Academic archaeology has focused on identifying location-focused properties based on different functional, chronological, and cultural types or components. Cultural resource management (CRM) in the United States and Canadian government agencies have used predictive models for cost assessment for developers to satisfy required archaeological surveys prior to ground disturbance. Archaeological heritage management (AHM) in parts of Europe and CRM in North America have focused on conservation and limiting costs, regardless of the nature of the resource (van Leusen et al. 2005, 27). It is not clear yet if these are appropriate uses of predictive models. Wheatley (1993), who originally was a proponent of predictive models, is adamantly against it now, especially for government agencies.

2.3.1. History of High Potential Models

Green (1973) performed a location analysis of Maya sites in northern Belize (then British Honduras). She identified soil types, vegetation, and distance from navigable water as the variables significantly associated with settlements. She produced a multivariate statistical model of site locations in areas with large tracts of arable land in proximity to trade routes.

In the United States during the late 1970s, the Bureau of Land Management (BLM), Forest Service, Corps of Engineers, Interagency Archaeological Services, and some State Historic Preservation Officers (SHPO) began to build high potential models. They sponsored surveys to develop the models and built models from existing survey data. The National Historic Preservation Act of 1966, amended in 1976 and 1980, signalled the importance of the growth of high potential modelling (Kohler 1988, 34). The subsequent high potential models utilised divergent methods and variables. Sebastian and Judge's 1988 volume entitled *Quantifying the Present and Predicting the Past: Theory, Method, and Application of Archaeological Predictive Modelling* developed out of this effort and serves as a key text in high potential modelling literature (Verhagen and Whitley 2012).

There have been significant advancements in site location modelling since the mid-1980s. Issues of modelling multiple site types, reconstructing the environment, and site sampling issues were identified as problems to be addressed. Advancements and adoption of GIS have allowed for new variables, new modelling approaches, and methods for evaluating model performance (Kvamme 2006, 21). GIS has also allowed novice researchers to develop simple models for site locations quickly and easily using new tools and software. Since the 1990s, there were a few publications each year regarding archaeological predictive modelling.

2.3.2. High Potential Models as Middle-Range Theory

Middle-range theory consists of the ideas, models, and other interpretive assumptions necessary to link low-level theory or low-level artefactual observations and high level or general theories (Bell 1994, 16). Binford (1977, 6) defines middle-range theory as '(a) how we get from contemporary facts to statements about the past, and (b) how we convert the observationally static facts of the archaeological record to statements of dynamics'. He saw these theories as a means to understand the processes responsible for change (Binford 1977, 7). Middle-range theory should be developed hand-in-hand with general theory because middle-range theory is designed to test general theories or upper-level theory (Binford 1977; Tschauner 1996; Raab and Goodyear 1984). Middle-range theory is the critical bridge between general theory and empirical data (Verhagen and Whitley 2012, 64).

Binford's view of middle-range theory is more limited than its use in sociology where it was originally developed (Raab and Goodyear 1984; Verhagen and Whitley 2012). It was intended to 'vary in levels of abstraction, to be flexible in seeking sources of working hypotheses, and to be aimed at accumulating a body of theory' (Raab and Goodyear 1984, 257). Application of Binford's middle-range theory largely has been reduced to experimental archaeology and ethnoarchaeology (Tschauner 1996). Raab and Goodyear (1984) advocate limiting or restricting middle-range theory development to empirical explorations.

Verhagen and Whitley (2012, 64–67) reduce the requirements for middle-range theory into four premises:

1. It is unambiguous;
2. It provides plausible cause and demonstrable effects not based on simple correlation;
3. It follows uniformitarian assumptions, and
4. It is independent of general or high-level theories.

The third premise may be difficult to apply to social theories because there may not be present-day correlates to use as benchmarks. Binford (1981) emphasises that middle-range theories should focus on process, and 'seems to have been unnecessarily restrictive in this respect [use of uniformitarianism], possibly because he envisaged middle-range theory to be used primarily for explaining site formation processes' (Verhagen and Whitley 2012, 67). In relation to the need for independence from general theories, though the middle-range theory is testing a high-level theory, it should be different or 'intellectually' independent of the high-level theory it is testing. To be a valid test, a theory cannot be tested by itself and thus has to have an independent theory to test it. A middle-range theory is this independent theory to test a high-level theory.

This research conforms to the premises defining middle-range theory. It is unambiguous because it postulates that, at times of lower sea levels, people would have lived on the continental shelf. The coastline along which they were living, fishing, hunting, and gathering would have moved progressively landward as sea level rose, and they would have maintained proximity to the coast and coastal resources. Secondly, there is a clear cause of change, late-Pleistocene/early-Holocene sea level rise. The effect of the change would be a landward movement in land-use and resource procurement locations. Thirdly, where people live in association to coastlines can be uniformitarian in nature. There are numerous modern and ancient examples including Carlson and Baichtal's (2007; R. J. Carlson and Baichtal 2015) record of site locations in relation to sea level in southeast Alaska during periods of higher sea level. Finally, this theory is independent of landscape theory, which is the specific high-level theory employed. Therefore, this middle-range theory meets Verhagen and Whitley's (2012, 64–67) premises.

2.3.3. Types of High Potential Models

Kohler (1988, 35) proposes three distinguishing dimensions to archaeological high potential models: level of measurement, procedural logic, and target context. He defines six types of models (Kohler 1988, 36):

1. Has no measurement, uses inductive logic and has a systemic context,
   - Ex. Binford's (1980) forager/collector model;
2. Has no measurement, uses inductive logic and has an analytic context,
   - Ex. Unqualified discussion of ancient land-use patterns;
3. Nominal/ordinal scale measurements, uses inductive logic, and has an analytic context,
   - Ex. Soil types associated with archaeological sites;
4. Nominal/ordinal scale measurements uses deductive logic and has a systemic context,
   - Ex. Using political boundaries for variable criteria;
5. Interval/ratio scale measurements uses inductive logic and has an analytic context,
   - Most CRM high potential models; and,
6. Interval/ratio scale measurements uses deductive logic and has a systemic context,
   - Models based on optimal foraging theory.

Model types vary from those with no measurement to models based on interval or ratio level measurements. Lack of measurements means researchers must rely upon qualitative statements, such as sites are frequently located near water. The intermediate scales for measurements are categorical or nominal variables, such as soil types, or ordinal level. This includes variables such as ranked resource lists. Interval or ratio level measurements include measurements like slope, distance to water, estimated net primary productivity, etc.

The target context refers to the 'theatre of operations' of the model (Kohler 1988, 35). Systemic context (Schiffer 1972) is the dynamic living systems observed by ethnographers and ethnoarchaeologists. The analytic or interpretive context begins when archaeologists start to sample materials or begin to observe their systemic context. The goal of locating archaeological sites implies a focus on analytic context (the archaeological site), but deductive models are concerned with systemic contexts (Kohler 1988). However, it is frequently unclear which context is being modelled despite critical differences in assumptions, approaches, difficulties, and outcomes (Kvamme 2006, 13).

Deductive models start with a theory about how people use a landscape and then deduce where archaeological materials would be located. Most CRM models have been inductive. Inductive models are also called inferential or empirical/correlative models based on their logical construct. These begin with survey data regarding the distribution of known archaeological sites and the site's relation to landscape features, usually environmental variables such as soil, slope, aspect, etc. The model then estimates the spatial distribution of the population of archaeological materials from the sample population. Inductive models work within the analytical context because they are attempting to control the post-depositional and depositional process that separates the analytic from the systemic contexts (Kohler 1988, 37). Explanatory models are neither inductive nor deductive, but instead build the bridge between systemic and analytical contexts. An explanatory inductive model would not have testable predictions, but an explanatory deductive model would result in predictions for analytical contexts. Thus, models are either inductive and analytic or deductive and systemic. Although others have created various schemes to classify models (Altschul 1988, 63), the majority are divided into inductive or deductive (Ejstrud 2003; Kvamme 2006; van Leusen et al. 2005; Verhagen and Whitley 2012; Kincaid 1988).

Kohler (1988, 52) recognises that deductive approaches are preferred for predicting the type and location of archaeological sites. A model's utility also can be judged by its ability to extend beyond common sense or causal explanations of site locations. Kvamme (2006, 11) echoes Kohler's comment but states that it was the researchers' processual framework, which was Sebastian and Judge's focus (1988): '[W]ho [the researchers and editors] informed us that only models generated through deductive reasoning were 'good' and potentially 'explanatory,' while models utilising statistical methods were not only 'inductive' (a bad word at the time), but "merely correlative" and incapable of explanatory insight' (Kvamme 2006, 9). Views have shifted since 1988. Inductive models tend to be based on statistics and deductive models are not popular due to their failure to apply statistics. Kvamme (2006, 12) explains how statistical models are a means to estimate appropriate weights for theoretically derived variables. This means using inductive modelling techniques to derive deductive variables. That is the method employed for this research. An example of this approach is Dalla Bona and Larcome's (1996) model for northwest Ontario, Canada, which is based on variables selected through ethnohistoric and contemporary native informant accounts.

The difference between data-driven models (inductive) and theory-driven models (deductive) is not universally recognised (Wheatley and Gillings 2002, 162). Models are neither purely deductive nor inductive (Kohler 1988, 52). Data is collected within a theoretical context and is thus theory-laden. Theories are based on empirical observations. Practically, it is impossible to implement a high potential model based entirely on either inductive or deductive methods (Wheatley and Gillings 2002, 162).

2.3.4. Testing Models

Testing includes two main aspects: consistency and evaluation (Verhagen and Whitley 2012, 84). All models, no matter how they are developed, need to be tested before they can be relied upon or operationalised. A model is operationalised if all the terms in the model are carefully defined and different people can make the same predictions using the same model. This also means that a model can be objectively and repeatedly mapped (Kohler 1988, 35).

Validation is 'the process of determining the accuracy and precision of the sample-based predictions either on internal criteria or on a portion of the target population not used in building the model' (Kohler and Parker 1986,

430). Validation tests the internal *consistency* of the model and the theory behind the model (Verhagen and Whitley 2012, 84; Ramenofsky and Steffen 1988, 9). There is little agreement about how to validate models and many are not validated. The best way to validate an archaeological predictive model is through field survey. This means that areas not previously surveyed can be tested for 'fit' between the expected and observed site patterns. If the desired fit is not achieved, the model goes through an iterative process whereby it is revised to incorporate new data and other variables (Kohler and Parker 1986, 430–32). Methods to achieve validity include resampling, such as cross-validation, jack-knifing, and bootstrapping. These can provide estimates of performance (Kohler 1988, 35; van Leusen et al. 2005, 34–35; Verhagen 2008).

The *evaluation* phase seeks to determine the predictive power of the model. It answers the question of how good the model is and its level of uncertainty (Verhagen and Whitley 2012, 84). This is often assessed using Kvamme's gain statistic (Kvamme 1988a).

$$\text{Gain} = 1 - \frac{\text{percentage of total area covered by model}}{\text{percentage of total sites within model area}} \quad [1]$$

This formula compares the per cent of the total study area to the total area of high probability. It is recommended that both training and testing data be utilised. Gain values approaching one have high predictive value, while values below 0.50 indicate that a model is not predictive. Results with negative gain values have reverse predictability or low potential areas have a greater chance of having archaeological sites than high potential areas (Mink II, Stokes, and Pollack 2006, 215; van Leusen et al. 2005, 33). The gain value is expressed as a decimal or converted into a per cent. It can be read as the probability to predict the location of a site. Generating the best possible model is a trade-off between accuracy and precision. Boundaries need to be defined to distinguish between high, medium, and low probability. Kvamme's gain can be used to determine these class boundaries to find the optimal threshold (Verhagen 2008, 286–87).

2.3.5. Issues with Predictive Models

Altschul and colleagues (2004) claim that many archaeologists believe that using a GIS equates to predictive modelling. However, others believe GIS is theoretically neutral (Wheatley 1993, 134). Arguments against predictive models focus on the issue that they cannot predict *all* archaeological sites. Other issues identified by Mehrer (2006, ix) include the perspective: [1] that site locations cannot be modelled because ancient cultures cannot be modelled; [2] that site locations cannot be model using known site locations because the population of known sites is biased by sampling errors; and, [3] that site models based on environmental factors are flawed because they are environmentally deterministic. Each of these criticisms is a generalisation about predictive modelling and it should be understood that a model is not trying to reconstruct all ancient culture, just the decisions underlying the selection of site locations. It also should be recognised that deductive models are not based on the location of known sites. Environmental variables provide independent statistical tests to evaluate objectively site location information and are valid variables as long as human choices are considered.

Kvamme's (2006, 7) table 1.1 provides a list of the archaeological, environmental, behavioural, and technical difficulties associated with archaeological predictive modelling. This includes the fact that many archaeological sites are buried/unknown and consequently their location cannot be used for modelling. He also notes that past environments were different from the present and may not be applicable for modelling. Finally, GIS data have insufficient resolution and poorly represent the real world. However, he counters this list of twenty-five problems with a list of three reasons why researchers can pursue archaeological location models:

1. Human behaviour is patterned with respect to natural and social environments;
2. Human environmental interactions can be observed from the spatial relationships between human residues (i.e., the archaeological record) and environmental features; and,
3. GIS provides a tool for mapping what is known (Kvamme 2006, 6).

Some of the difficulties, such as poor resolutions of GIS are beginning to see improvements. GIS has a theoretical background (discussed in the next section) that has to be incorporated and included when developing any model. GIS make assumptions about the environment that can affect how the model is produced, such as the scale a variable is recorded or displayed.

With land-use predictive models, there are two types of errors. The first is where a site is predicted and none exists. This can be termed a 'wasteful errors', errors of commission, or a Type II error in statistics. Type II errors occur when the null hypothesis is accepted, and it is actually false. The second type is where a location is predicted not to have a site and it does. This type of error is a Type I, errors of omission, or 'gross error'. Type I errors occur when the null hypothesis is rejected and it is, in fact, true (Altschul 1988, 62). Ideally, a model should reduce both types of errors. For archaeology, especially CRM, it is costlier to make gross errors than wasteful errors.

2.3.5.1. Site Types

Kvamme's (2006) niche model explanation clearly articulates the need to model activity areas, not all site types. Site inventories often include scatterings of only a handful of lithics, petroglyphs, and culturally modified trees (CMTs). These surface reconnaissance inventories do not allow for reliable reconstructions of other activities that may have contributed to the formation of archaeological sites. Specifically lacking are dates of use for some of these locations.

Modelling specific activity types requires creating a subset of data. These subsets often are small and consequently can be so few as to be meaningless statistically. The use of all possible site types is the reason why most models remain dichotomous, indicating either the presence or the absence of sites. Despite this limitation, simple presence-absence models can be very useful (Kincaid 1988, 561; Kohler and Parker 1986, 414–15). For a meaningful model of specific site types, each class of site included must be a statistically significant site population. Methods to increase site counts or sample size include additional fieldwork, increased funding, retrieval of larger samples of artefacts, and more efficient analysis methods. Larger numbers of sites attributable to specific activities can improve theory related to site function and location (Kohler and Parker 1986, 414; Kvamme 2006, 18).

2.3.5.2. Archaeological Site Sample

When creating an inductive site location model, usually the existing archaeological site record is utilised. However, many of these sites were recorded prior to geographic positioning systems (GPS) and the data are flawed because many site locations were not accurately recorded. Even using GPS, recording accuracy can be off by several tens of meters and other errors can occur during data entry. Often polygons are actually provided to the recording agency rather than point specific recordings. Frequently, it is not specified whether the site location was recorded as either points or polygons. As a result, often the polygon centre is used and there can be significant differences between the centres of the polygon and the site's actual location. These 'points' are used for spatial correlation with environmental variables. However, the amount of error in their locations can result in significant statistical outliers biasing the sample population. These inaccuracies can result in flawed modelling (Altschul and Nagle 1988; Kvamme 2006, 20, 1988b). The high resolution of the model (two-meter) means that sites that are more than two m from their actual location can affect the statistics used for the inductive portion of the model. These can skew the values further from the real-world results. Some sites are not even on land, as per the USGS topographic maps; this is a significant error that propagates throughout the model. Since archaeological site locations are also used as an input in the model, the inaccurate location then also has a higher probability of locating other sites in the wrong location. These errors in the location of sites also affect testing the model (chapter six) as the site locations were used to assess the evaluation of the model.

Archaeological site inventories also are biased because archaeologists have a tendency to look for and discover sites where they believe they should be (i.e. along rivers or lakes). They can also be biased by preferentially selecting locations that are easier to access for an archaeological survey (i.e. near roads and towns). Sites inventories also result from CRM surveys of non-random locations selected for development projects. Finally, larger sites and land-use (such as those with mounds, earthworks, or structures) have a higher archaeological visibility (Altschul and Nagle 1988; Kvamme 2006, 1988a) and tend to be over-represented in relation to other site types. These biases than are incorporated into locational models based on existing site inventories. Hence, the assumption that inductive models can only predict locations based on the known site locations is not correct, and a combination of deductive and inductive modelling techniques are necessary to achieve greater accuracy.

Some modelling projects employ random sampling methods and pedestrian surveys in an attempt to create unbiased samples for model development. Usually, probabilistic sampling strategies are utilised to minimise cost. This sampling strategy can reduce estimates of site variability and result in a lack of independence between data elements. Consequently, these data can become spatially autocorrelated (Kohler and Parker 1986, 408–10; Kvamme 2006). This method was not viable for this underwater project.

Despite these problems in using existing archaeological site data, archaeological predictive models have been developed and successfully employed all over the world. Ejstrud (2003) utilised the Danish National Sites and Monuments record for predictive models for Denmark. Kvamme (1988a) provides multiple statistical techniques for accounting for existing archaeological site data.

2.3.5.3. Palaeoenvironments

Models often employ modern environmental variables without acknowledging that modern conditions can be significantly different from those of the past (Kvamme 2006, 16). In southeast Alaska, sea level change prior to 5,800 cal BP (5,000 ^{14}C years) transformed the coastal zone, the area, and configuration of the available land, and the distribution of biotic resources. The total difference in sea level since the LGM was -180 m. During the LGM sea level was 165 m lower than modern sea level. Then sea level was 15 m higher than modern sea level (Baichtal and Carlson 2010; R. J. Carlson and Baichtal 2015). This means that models incorporating land/sea level relationships for the last 5,000 years reliably could utilise modern variables related to sea level, but models for periods prior to that time require palaeoenvironmental reconstruction. The 5,000-year period is only valid for southeast Alaska and needs to be determined for each region based on its sea level reconstruction.

Researchers use modern data for modelling because it is readily available and the error is often quantifiable (Deeben et al. 1997, 81–84; Kondo and Oguchi 2012, 5–6; van Leusen et al. 2005, 56–57; Wheatley 1993, 135). Terrain forms, such as mountains and valleys, do not change significantly through archaeological time scales. Rivers and streams will meander within valleys. Plant communities will migrate, or become extinct, with climatic changes. Although the shape and slope of a landscape may not change significantly, the scale of changes over time must be understood and regional palaeoenvironments reconstruction may be required as a first step in site location modelling (Kvamme 2006, 19; Kondo and Oguchi 2012, 5–6; van Leusen et al. 2005, 56–57; Wheatley 1993, 135). Palaeoenvironmental reconstructions rarely have been undertaken for archaeological site modelling because they are difficult and often require broad generalisations. Errors in the original data used for palaeoenvironmental reconstruction can result in 'huge errors owing to their imprecision' (Kvamme 2006, 19). However, reconstructions are still better than using modern data when researching the distant past (Deeben et al. 1997, 81–84; Kondo and Oguchi 2012, 5–6; van Leusen et al. 2005, 56–57; Wheatley 1993, 135). The reconstruction for of the palaeoenvironment for southeast Alaska is reasonable based on testing against National Oceanographic and Atmospheric Association (NOAA) charts and multibeam data. There are areas of the study region that have poor resolution; these areas are not within the survey area. Higher resolution would be beneficial, but the resolution is adequate for locating the variables used in this research.

2.3.5.4. Scale

A fundamental question for predictive modelling is geographic and temporal scale, or resolution. Scale in the geographic sense can be divided into three categories: macro-, meso-, and micro-scale (Saile and Lorz 2005). Archaeological research moves between scales of analyses, from the unit of an artefact to features to sites to landscapes (Altschul 1988, 71; van Leusen et al. 2005, 58–59; Verhagen and Whitley 2012, 89). The scale of analysis and the scale of results is not always the same, and scales need to be reported. Geographers refer to this problem as 'ecological fallacy' which is the danger of erroneous extrapolation from one scale to another (Lock and Harris 2006, 4–5; Verhagen and Whitley 2012, 90).

Predictive modelling requires a uniform scale of analysis, in terms of both resolution and geographic extent. The analysis must be conducted and interpreted at a defined resolution, which usually is described as a certain number of square meters or square miles for each unit. Whatever scale is used, the environmental and social variables utilised by the model should then be reduced to the same scale. This means that for a one-km^2 grid, any geographic feature smaller than one km would not necessarily be identifiable. The one-km^2 grid would have an average value for the terrain, water resources, subsistence resources, or any other variable (Altschul 1988, 71). Scales such as one-km^2 are not reasonable units for archaeological analysis and this type of large-scale analyses generally is conducted due to lack of computing power or poor data resolution. The resolution of the model needs to be at less than the feature or site that is being modelled. The scale of predictive modelling analyses should continue to decrease to the scale of an archaeological feature, as these factors improve. Saile and Lorz (2005) describe scale on a continuum with the largest scale only suitable for a general (large-scale) approach. The middle, or optimal, is an appropriate scale for the research objectives. Finally, there is a scale that is too small, that of an individual case (or artefact). The scale of analysis for this study originally aimed to be that of an archaeological feature. Unfortunately, computer limitations restricted the digital elevation model (DEM) to two-meter, which is too large to identify individual features. The goal was for a sub-meter DEM and one-meter model. The model scale of two-meter is ideal for identifying landforms where archaeological sites are probable. It can identify the features being mapped by the model variables, such as terraces and flat areas near freshwater.

The exact geographic extent of a study is often determined based on outside influences, such as political boundaries or available data. This imposes limitations on spatial patterns (Lock and Harris 2006; Verhagen and Whitley 2012, 90). The literature rarely includes descriptions of how or why a study area or extent is selected (Wheatley 1993, 136). Maschner (1996) recommends data exploration to help eliminate, or limit, both the extent and scale of analysis. However, the scale of cultural perception needs to be incorporated into the analysis (Wheatley 1993, 135). The 'scale of cultural perception' is any scale a person can view the landscape, i.e. how far can you see in a dense forest.

Temporal scale is also important to predictive modelling. Grouping or splitting temporal units can cause models to ignore important events such as significant periods of climatic change (Kohler and Parker 1986, 406). Sites 'as they are observed by archaeologists are created by the act of observation at a particular point in time' (Dunnell 1992, 27). Dunnell (1992, 28) goes on to observe that sites and artefacts acquire their attributes over an amount of time. Site activities can be distinguished based on their duration (Wandsnider 1992b, 257).

Chronological schemes are historical constructions rooted in evolutionary theory that has been developed beginning in the 19th century (Peeters 2005, 152). These schemes (for example, Stone Age, Bronze Age, and Iron Age) can still be applicable in predictive modelling. The schemes must be evaluated statistically for similarity and/or differences. Absolute dates (such as radiometric methods or tree-ring dating) are more complex and consequently more difficult to apply. The temporal units of analyses must also be defined and justified based on the goals of the predictive model and research question(s) (Peeters 2005). For this research, calendar years were used to assist with interpretation and analysis. The scale of 500 years is the finest resolution that could be determined with the low resolution of the sea level reconstruction and the few archaeological sites from these periods. Finally, the lower limit of 10,500 cal BP was selected because it was the closest value to when sea level rose above modern values. The maximum limit of 16,000 cal BP was selected because that is the farthest back the sea level reconstruction extents currently and it corresponds to when the region is thought to have become deglaciated.

2.3.6. Examples of Predictive Models

Barber and Roberts (1979; Kohler 1988, 46–47) developed a model for the continental shelf from the Bay of Fundy in Maine to Cape Hatteras in Northern Carolina. This area is currently submerged but was exposed at or after 18,000 cal BP. The authors used both inductive and deductive approaches to estimate site types and densities on the continental shelf. The deductive model was based on optimal foraging theory. Barber and Roberts simplified the model by identifying the potential for food resources, as primary or secondary. Primary resources were in the immediate vicinity of a site, such as freshwater. Secondary resources were further afield from the site, such as larger prey animals. Although they used values that were experimentally estimated, this approach minimised the need for accurate palaeoecological reconstructions. This is a rare and early example of a North American predictive model to locate ancient sites on the continental shelf.

In the early 1990s, the Ontario Ministry of Natural Resources (MNR) in Canada became responsible for identifying and protecting cultural resources through forest management planning. This led to a 'first generation' predictive model based on ethnographic and environmental variables by the Centre for Archaeological Resource Prediction (CARP) (Dalla Bona 1994; Dalla Bona and Larcombe 1996; Hamilton and Larcombe 1994; Dalla Bona 2003). It took three years to develop the model, which focused on the Boreal forest ecotone of northwest Ontario. The model incorporated both an inductive and deductive approach. The ethnographic literature was reviewed to help define subarctic foraging land-use systems, prey selection, and seasonal rounds for the deductive portion of the model. Northwest Ontario had not had a significant amount of archaeological survey, and the research team conducted extensive surveys to generate a database for inductive modelling (Hamilton and Larcombe 1994, 59). To generate the model, they utilised a weighted overlay method with the weights of the environmental and social variables determined based on statistical analysis of existing sites. The modelling was followed by three additional years of pilot projects to expand the model's applicability to other areas of Ontario (Dalla Bona 2003). This model's methods and weighting system were utilised to guide this research.

Mackie and Sumpter (2005) utilised shoreline intricacy to analyse site distribution patterns on Haida Gwaii. They defined four categories of shoreline complexity: linear, sinuous, elaborate, and intricate, in increasing complexity. Their premise was that greater shoreline length near a site would allow for more readily accessible intertidal and subtidal resources, and theoretically greater biodiversity and increased opportunity for subsistence activities (A. P. Mackie and Sumpter 2005, 350–51). They conclude that early sites (9400-9500 C^{14} years) existed on more 'elaborate coasts'. Late sites (2000 C^{14} years to modern) were distributed more evenly between the shoreline intricacy categories. Mackie and Sumpter utilised habitat and distance from known marine resources as a key parameter. The model uses the modern landscape and modern resource distributions. They never explain why or how the modern landscape is appropriate for this research or justify their choice. Their study employs a landscape approach and provides useful information that helps explain ancient site location choices. It also introduces the concept of coastal sinuosity or intricacy as an archaeological coastal land-use-modelling variable.

Maschner (1992) developed a predictive model for Kuiu Island, north of Keku Strait in southeast Alaska. In three field seasons, Maschner found 148 sites bringing the total to 155 for his study area. He worked from the high-water line to the base of the mountains (or until it was too steep to survey). Maschner and Stein (1995) applied logistic regression to the survey results and a random sample of non-site locations to determine the environmental requirements for ancient decision-making. For camps and villages, the characteristics that Maschner and Stein (1995) determined to be important are southern exposure, a good beach for canoe landing, and protection from storms. Some archaeological sites did not have access to freshwater at the time the survey was conducted.

Recent research by a Tongass National Forest archaeologist (Risa Carlson) and geologist (James Baichtal) has identified the 10,600 to 7,800 cal BP (9,400 to 7,000 ^{14}C years) shorelines. These shorelines are elevated above the modern shoreline by up to 20 meters. Carlson and Baichtal (2015) have developed a high potential model to locate archaeological sites within this time-period in the Tongass National Forest for several areas on the west side of Prince of Wales Island (POWI). The model has high predictive validity, as they have located over 70 archaeological sites from modern sea level up to 32 meter above zero tide. Carlson and Baichtal (2015; Baichtal and Carlson 2010) focused their tests in the NW corner of POWI, adjacent to Shakan Bay. Baichtal has proposed that by 9,000 BP, this area was an interaction corridor and that people were on the coast prior to 9,000 BP. This supports the hypothesis that rising sea levels submerged the evidence of earlier occupations.

2.4. Geographic Information Systems (GIS)

GIS is a tool that researchers use to investigate, store, analyse, and visualise spatial phenomena such as artefact and site distributions. It is both a database system with spatial referencing and a set of operations for working or analysing data (Wheatley and Gillings 2002, 9). Some practitioners see GIS as a science, not simply a tool. Like high potential modelling, others see GIS as theoretically neutral. Nevertheless, the questions, methods, interpretations, and results that the researcher generates from the program can be tested scientifically. There are many theoretical aspects to GIS and these theoretical frameworks vary based on the program used and the research goals.

As one of the low-level theories employed by this research, GIS provides a tool and methodological framework where both landscape archaeology and land-use modelling can be employed and tested. The analytical units are actually created and measured in the GIS. The final outputs for this model are created in GIS and then developed into maps that also were created in GIS.

2.4.1. History of GIS

The development of GIS for archaeology is dominated by influences from outside disciplines. Originally developed by geographers for map classifications, the first operational integrated GIS was created by the Canada Land Inventory for digitising maps (Lock and Harris 1992). As a tool for archaeology, GIS allows complex and time-consuming analyses to be conducted in less time and opens these analyses to non-specialist users.

GIS has been in existence since the 1960s. The Canada Land Inventory's operational integrated GIS was developed by 1972. It was not until the early 1980s that GIS began to be used by archaeologists. By the mid-1980s, user-friendly GIS programs were being developed, such as ESRI's ArcGIS and GRASS, including modules specifically for archaeologists. The movements toward the use of archaeological GIS originally were focused in North America, and there is still expansion and growing awareness of the potential of GIS in the United States (Lock and Harris 1992; Mehrer 2006, ix). 'In less than ten years, GIS has progressed from exotic experimental tools, requiring costly workstations and only available to a few specialist researchers, to widely available technological platforms for the routine analysis of spatial information' (Wheatley and Gillings 2002, 1).

GIS is becoming one of the most widely used analytical applications for archaeology. Since 1991, there have been advances in both the complexity of GIS software and the design of applications. This can be related to an increase in the capabilities of both hardware and software. This increase in analytical capacity and capabilities is unquestionably one of the most significant advancements in archaeological modelling (Daly and Lock 1999).

2.4.2. GIS Programs

There were three main programs utilised for this research: ArcGIS and ArcPro from ESRI (Environmental Systems Research Institute), GRASS (Geographic Resource Analysis Support System) (GRASS Development Team 2015), and QPS Fledermaus. ArcGIS is proprietary software that costs individuals thousands of dollars for licensed use. Its popularity is based on university and government subscription to its database and file formats. In North America, ArcGIS is the standard platform taught for GIS and utilised by the United States government. In Europe, as a cost cutting measure, governments and universities are encouraged, and often required, to utilise open source platforms such as GRASS, QGIS (Quantum GIS), or GVsig. These open source platforms are free and the source code is available for those capable of correcting errors or developing new tools. This divide between proprietary and open source platforms has caused significant difference in the tools and methodologies used in North America and Europe. QPS Fledermaus is another GIS platform specifically designed for maritime resources and like ArcGIS, is proprietary software.

Data can be characterised as raster, vector, TIN (triangulated irregular network), or terrain models. The last two, TIN and terrain models, have limited functionality (ESRI 2001). Vectors are points, lines, and polygons. They have information attributed to each piece. Data are contained in vertices that are similar to points. For lines and polygons, straight lines join these vertices and this information is not accessible to the user except in specific edit modes. Raster data contain an array of rectangular cells. The geographic information then is assigned to each cell. The resolution of a raster file is based on the cell size (Longley et al. 2005, 74–75).

Each operation in a GIS, either statistical or manipulative, has a theoretical framework. The key operation for this research is ArcGIS's weighted overlay. To utilise this tool, all data must be reclassified into a nominal numbering system. After each raster file is ranked and weighted, the raster is multiplied by its weight. The weighted rasters are added to create the final output (ESRI 2011). Other GIS tools are explained in the methods sections.

2.5. Underwater Archaeology

The ocean covers 72 per cent of the earth's surface (Ballard 2008, ix). Sea level has risen by approximately 120 meters globally since the LGM, and the area flooded represents three per cent of the earth's land or four million km^2 (Coleman 2008, 178). As a result of post-Pleistocene sea level rise, archaeological remains located on the world's continental shelves are underwater. Underwater archaeology investigates sites that are submerged or located in the intertidal zone. As a sub-discipline within archaeology, varieties of terms have been used to describe the discipline including maritime, marine, nautical, oceanographic, and hydro-archaeology. Each term limits the scope of research. Maritime Archaeology investigates the maritime cultural traditions, including shipwrecks, graves, harbours, ship related technologies, and shipbuilding facilities (Croome 2004; Forrest 2002). Underwater archaeologists investigate palaeo-landscapes and other submerged environments (Flatman 2003; Parker 1999). Muckelroy's (1978, 4) definition of maritime archaeology is the most recognised in the sub-discipline. He defines it as 'the scientific study of the material remains of man and his activities on the sea'. Though Muckelroy allows for the incorporation of aspects of nautical and underwater archaeology, he excludes gravesites and submerged landscapes. Wetland environments (Coles 1998, 2004; Flatman 2011; Lock 2004; Purdy 1991), shipwrecks (Gould 2000; McGrail 1981; Watson 1983), and inland-submerged environments, such as lakes and rivers (Ford 2011b; Halligan 2011; Mazurkevich and Dolbounova 2011) were not included here as they are outside the scope of this research.

Underwater archaeological sites can be divided into four categories. The first are shipwrecks. The second are shrines or places of offerings, such as the cenotes of Mesoamerica. Third are items discarded or lost, refuse sites. These are most common when material purposefully is discarded into the water. Finally, there are submerged occupation sites (Goggin 1960; Ruppe 1988; Faught 2004; Cook Hale and Garrison 2017) which are the focus of this research.

Investigation and artefact retrieval from shipwrecks has occurred throughout history (Navy 1998). However, underwater archaeology as an academic discipline came into existence with George Bass' 1960 excavation off the coast of Turkey. Bass' excavation of a Bronze Age shipwreck was the first that met the professional criteria and standards of excavation and recording in archaeology. The scientific standards of underwater archaeology are the same as terrestrial archaeology (Bowens 2009, 6; Dean et al. 1992, 20). In 1854, the Germans and Swiss excavated lake dwelling sites when they were exposed due to low water levels (Christensen 1997). The Fos is a partially submerged ancient village that was one of the first prehistoric underwater site that was excavated in 1964 (Baucaire 1998). In North America, underwater archaeology began in the 1930s with the National Park Services excavation in Colonial National Historical Park, Virginia (G. Fischer 1974).

Technological developments in marine geophysics have increased the depth range for investigating submerged locations (Coleman and Ballard 2008; J. Green 2003). Theoretically, there should be no difference between geophysical surveying on land or underwater, except for logistical issues (Coleman and Ballard 2008, 4; J. Green and Gainsford 2003). Survey methods include bathymetric mapping, side scan sonar, magnetometer surveys, high-resolution reflection surveys such as sub-bottom profiling, and visual imaging surveys using remotely operated vehicles (ROV). These technologies will be discussed further in the surveying chapter (chapter 7: Field Testing Results).

Site preservation is a key issue for underwater archaeology. The anaerobic environment encountered centimetres below the seafloor provides excellent preservation of organic materials (Peterson 1974). Northern oceans, including the Arctic and North Pacific, provide a cold environment that further decreases microbial activity and increases organic preservation. In addition, preservation is improved by the saturation of the organic artefacts by water (A. Fischer 1995a). Although marine transgression can be destructive to archaeological sites, Coleman and Ballard (2008, 6) describe how rapid burial provides an improved chance for site preservation. They describe lithic artefacts, kitchen middens, gravesites, stone foundations, pottery, hearths, postholes, and other cultural features as the types of cultural remains that can survive marine transgression. These 'non-delicate' remains are detectable evidence of former human occupation and clearly identify a site, providing evidence for archaeological analyses.

Kilgii Gwaay is an intertidal site on the south end of Haida Gwaii, British Columbia, Canada that was occupied approximately 10,700 cal BP. It includes delicate basket remains despite the site having survived at least one marine transgression and subsequent marine regression (Fedje et al. 2001; Fedje, Mackie, et al. 2005; Q. Mackie et al. 2011). Finally, the existence of palaeontological remains located in the Bering Sea implies, by analogy, archaeological remains can survive marine transgression in this region (Dixon 1983; Dixon and Monteleone 2014). Masters (1983, 211) notes the present sea level has not adversely affected the underwater Solana Beach reef site, California, located in a small cove. She postulates that some coastal sites may have survived marine transgressions in similar low energy environments. This is a point taken up by Garrison and colleagues (2016) when they argue that recent finds have invalidated Meighan's (1983) concluded that it was unlikely that 'Early Man' sites would be located underwater.

Submerged archaeological sites usually are found by accident by divers, dredging, and fisherman. Few have been located by systematic archaeological survey (Flatman and Evans 2014). Underwater archaeology can be very costly and difficult. It has been poorly sponsored by funding agencies (Coleman 2008, 177), despite the fact that submerged landscapes are being eroded due to the death of eelgrass and similar vegetation around the world because of pollution. Once eelgrass or similar coastal vegetation dies, the underlying archaeological sites start to erode and can be completely destroyed (A. Fischer 2011, 303–6). The poor funding is likely linked to the exploratory nature of this type of research and the possibility that no archaeological sites (or non-sites) will be preserved or located within the survey areas.

2.5.1. Danish Model

In Denmark, laws protect submerged land-use and it is illegal to harm or remove the archaeological remains without National Forest and Nature Agency approval (Lund 1995; A. Fischer 1995b). As a result of post-LGM sea level rise, many Palaeolithic and Neolithic sites in Denmark are now underwater. Many of the Early Mesolithic sites of the Maglemose culture and the Middle and Late Mesolithic cultures of Kongemose and Ertebølle are underwater (A. Fischer 1995a).

Fischer (A. Fischer 1995a, 373–74, 2004, 27–32) presents a fishing site location model that is based on archaeological and ethnographic data from Roskilde and Karrenæk Fjords. Local fisherman from Karrenæk Fjord, who still practice traditional fjord fishing, provided the topographic characteristics of the best fishing locations.

They are situated at the mouths of streams, where fjords narrow, or offshore of small island and peninsulas near sloping bottoms. These topographic patterns are seen throughout southern Scandinavia and other places where maritime cultures are common. 'The great majority of the coastal settlements from the Ertebølle Culture in Denmark and Scania have been situated immediately by the shore, at places that would have been ideal for fishing with traditional stationary gear' (A. Fischer 1995a, 374). The model works well for Kongemose and Ertebølle cultures (A. Fischer 1995a). This is a simple landscape model. This would be an example of Kohler's (1988, 35) type one model, where there are no measurements, uses inductive logic, and has a systemic context.

Benjamin (2010) outlines the steps required to develop the southern Scandinavian model for submerged ancient site discovery, referred to as the 'Danish Model'. Phase I is regional familiarisation, including maps, archaeological site types and cultures. Phase II is familiarisation of the ethnographic literature. Phase III is location plotting of known sites. Phase IV is geophysical survey of high potential locations. Phase V is investigating possible sites using divers. Finally, Phase VI is post-fieldwork analysis, interpretation, and dissemination (Benjamin 2010, 258). Ford and Halligan (2010, 278) propose a slight variation of this model because some areas are too deep for divers. They amended Phase V to include 'with divers when feasible'. This model is similar to the procedural steps used for this study at a high level. Though referred to as the 'Danish Model', this is essentially the basic model for most archaeological research or fieldwork, though diving is replaced by field survey or generalised.

2.5.2. Underwater Archaeological Projects

Originally, the majority of underwater archaeology that occurred since Bass' excavation in the 1950s has been focused on shipwrecks and coastal structures including harbours and piers. More recently, there has been a significant increase in the number of prehistoric submerged landscape surveys and excavations (Bailey and Flemming 2008; Sturt et al. 2018; Flemming 2017). There continues to be extensive work in Europe. Project areas include Danish, Doggerland, and the Norwegian projects in the North Sea (Gaffney and Thomson 2007; Gaffney et al. 2017; Bryn, Jasinski, and Soreide 2007; Gljjrstad, Gundersen, and Frode 2017; Coles 1998, 2000; Glørstad 2014; Ch'ng et al. 2011; A. Fischer 1995a; Benjamin 2010; A. Fischer 2004; Bjerck 1995), the Solent (Momber 2000, 2011), and a Middle Palaeolithic site at La Mondrée in Normandy (Cliquet et al. 2011), to name and reference but a few of the global projects. There have been several new edited volumes reviewing coastal and continental shelf archaeology including *Trekking the Shore: Changing Coastlines and the Antiquity of Coastal Settlement* edited by N. Bicho, J. Haws, and L. Davis in 2011, *Prehistoric Archaeology on the Continental Shelf: A global review* edited by A. Evans, J. Flatman, and N. Flemming in 2014, *Geology and Archaeology: Submerged Landscapes of the Continental Shelf* edited by J. Harff, G. Bailey, and F. Luth in 2016, and *Under the Sea: Archaeology and Palaeolandscapes of the Continental Shelf* edited by G. Bailey, J. Harff, and D. Sakellariou in 2017. Again, this is not an exhaustive list but an example of how significantly submerged landscape research has grown in the past few decades.

What follows is a brief review of submerged archaeological projects in coastal environments in North America. These projects are described in geographic orientation circling counter clockwise from the northwest to the northeast. The number of these projects continues to grow, and this list is only a snapshot of what has and is being done around the continent. Halligan and colleagues (2016) work in the Page-Larson site is a riverine site, and thus not included in this review, but still requires a mention due to the age (14,550 cal BP) and submerged situation.

Dixon (1979) conducted a marine geophysical survey off St. George Island in the Bering Sea. He developed a high potential model for archaeological sites based primarily on bathymetric data, palaeoenvironmental reconstructions, and sea level history. At this time, there was not a bathymetric map for the Bering Sea and the research was conducted prior to authorise non-military use of GPS. The project attempted to identify areas of high archaeological potential based on high latitude hunter-gatherer subsistence and land-use patterns. Due to high winds and seas from a Bering Sea storm, the survey was conducted on the lee side of St. George Island, which was an area that had not been identified to be high potential for archaeological site discovery. The survey was conducted using a hull-mounted sub-bottom profiler, side scan sonar, and proton magnetometer. On-shore radio navigation stations established at prominent locations on St. George Island were used to achieve navigational accuracy of ± 50 feet (15 m). Though no archaeological remains were identified, this survey demonstrated the feasibility of searching in high latitudes for submerged archaeological sites (Dixon 1979, 1989; Dixon and Monteleone 2014).

In western Hecate Strait off the coast of Haida Gwaii, 10 km² of the seafloor in Juan Perez sound were mapped in 1997 using high-resolution multibeam sonar, side scan sonar, and an ROV (Fedje and Josenhans 2000). The sonar imagery revealed landscape dominated by alluvial fans, deltaic plains, and a meandering and migrating river system. In 1998 and 1999, Fedje and Josenhans (2000) focused on targets from the DEM generated from the multibeam survey to concentrate investigation for archaeological sites and to facilitate dating late Pleistocene shorelines. They recovered and hydraulically screened sediment using a clamshell grab sampler and screens. Within these sediments, they discovered buried wood and other evidence of the ancient terrestrial environment. They made the significant discovery of a retouched basalt blade-like flake from sediment recovered from a depth of 53 meters in Werner Bay. They concluded that this artefact was probably deposited on the continental shelf at a time when sea level was lower (Fedje and Josenhans 2000; Josenhans et al. 1997). In 2014, an autonomous underwater vehicle with a side scan sonar surveyed almost seven km^2 and identified stone alignments and house depressions in this same area (Q. Mackie, Fedje, and McLaren 2018, 79).

In Montague Harbour, Gulf Islands, British Columbia, Canada, Norman Easton conducted four years of underwater excavation. 'The goal of the research is [was] to develop an in-depth assessment of the potential, distribution, and nature of inundated cultural deposits within the Montague Harbour basin, a well-protected tidal embayment' (Easton 1993, 3). The midden that was excavated was located with a sub-bottom profiler, cored, and then excavated (Easton and Moore 1991). Artefacts recovered include an antler harpoon point from below sediments dating to 6,800 cal BP, non-benthic faunal, unifacial flakes, other flaked tools, and a stone disk bead. Easton (1993, 18) notes the presence of bioturbation in the underwater excavated locations but indicates that it is restricted to specific depositional zones. Retaining boxes made of metal were used to excavate to a depth of 2.5 m by divers. The Montague Harbour excavation was a Charles phase (5800-3500 cal BP) or earlier site (Easton 1993).

Marine geophysical and diver surveys were conducted off the coast of California's Northern Channel Island of Santarosae. Santarosae is the island that formed during lower sea level when San Miguel, Santa Rosa, Santa Cruz, and Anacapa Islands were joined as single land mass. Watts and colleagues (2011) developed palaeoenvironmental reconstructions to direct their field survey that led them to select areas near terrestrial Terminal Pleistocene shell middens for underwater survey. In 2008, they used diver survey to investigate high potential areas, and in 2009, they incorporated side scan sonar with the diver surveys. They recorded video and still images of the seafloor. In 2009, a chert cobble showing possible flake removal scars was identified but was left in situ (Watts, Fulfrost, and Erlandson 2011, 22).

In 1981, Muche (1998, 254) presented a model to locate archaeological sites in the Santa Monica Bay region of California at the Tenth Conference on Underwater Archaeology in San Marino California. He considered eight factors: [1] the ecological setting of known land sites; [2] geographic features; [3] isolated find reports; [4] water depth-time correlation; [5] indications of a pre-existing water supply; [6] ethnographic and historical documentation, especially from early Spanish records; [7] palaeontological evidence; and, [8] palaeoclimate reconstructions. A probability factor from one to ten was assigned to locations based on assessment of these eight factors. Eleven-out of 24 locations or 32.4 per cent of the locations were designated as high potential. Correlation of submerged geomorphic features produced an additional seven high probability areas. These 18 locations were field-tested and 13 had archaeological remains. Artefacts include mortars or bowls, pestles, points, drills, scrapers, and shell beads. Based on local sea level reconstructions, these sites were occupied 3000 to 11,000 years ago with an error of approximately 1000 years based on unstable geological conditions.

Patricia Masters (1983) conducted diver survey and located 22 new underwater sites in the San Diego County area including twelve previously recorded sites. Four were remnants of shell fishing. Two were tool manufacture sites. Fourteen of the sites were between 12 and 19 m depth. Others were intertidal.

Bahía Ballena is a submerged landscape off Isla Espírtu Santo in the southern Gulf of California. Faught and Gusick (2011) report a bathymetric survey conducted over an area of 30 km^2 with a maximum depth of 140 m. A total of 2145 depth readings were recorded by hand whenever the depth changed by more than 1.5 m. A DEM was created using the Inverse Distance Weight (IDW) in ArcGIS 9. The bathymetry data were merged with Shuttle Radar Topography Mission (SRTM) digital data to create a 12,000 cal BP landscape. Survey was focused on nearshore areas, shallow enough for SCUBA investigation and safe for divers. After initial survey, the research team focused on four submerged rock shelters, the areas outside each rock shelter, an extensive rock outcrop, and adjacent steep slope. Due to tidal action, it was difficult to distinguish between culturally and non-culturally modified flakes. Hand fanning was used for test pits to a depth of 80 cm (Faught and Gusick 2011; Guisck and Faught 2011).

The Gulf of Mexico had CRM reports from Coastal Environments, Inc. (Coastal Environments 1977) that mapped probability areas for submerged historic and ancient sites. Michael Faught conducted his dissertation research in the Big Ben of Florida investigating site presence and preservation using multibeam sonar and diver survey. Chipped stone artefacts, some characteristic of the mid-Holocene, were located along submerged river channel margins of the Aucilla River (Dunbar, Webb, and Faught 1991; Faught and Gusick 2011). Adovasio and Hemmings were funded by NOAA for a multi-year project to remotely sense and investigate with divers, sites in the Big Ben area. Florida is the only state that requires sub-bottom profiling for underwater CRM (Faught and Gusick 2011). Coastal Environments, Inc. has continued their research funded by the Mineral Management Service (MMS, now the Bureau of Ocean Energy Management (BOEM)). They updated their original 1977 manuscript in the 1980s and work continues within the region for oil and gas exploration and exploration of possible shell middens (Faught and Gusick 2011). Pearson (et al. 1986) conducted sub-bottom profiling and vibracoring of sites offshore from Louisiana and Texas along the Palaeo-Sabine River. Two probable archaeological sites were identified. The first was a quarry that included fine lithic debris, burnt bone, organic, and phosphates dating to 8850 cal BP. The second was a concentration of shells thought to be a midden.

Survey and analysis have been conducted at the Gray Reef Marine Sanctuary since 1996. Three artefacts have been recovered during diver survey including a stone projectile point, a flake scraper, and a bone or antler ata-atl hook (Garrison et al. 2016). This project continues to narrow the areas for diver search following Benjamin's 'Danish Model' and spatial statistics (Cook Hale and Garrison 2017). They are looking for areas of higher ground with flat areas 'behind' or landward of them. In this region, the sea level was significantly lower until 5800 cal BP (Garrison et al. 2016).

Stanford and colleagues (2014) have been investigating late Pleistocene projectile points found on the continental shelf near Chesapeake Bay. This includes the Cinmar site, which was a dredge find. This project is conducting coastal research and palaeoenvironmental reconstruction to identify where fishermen have located points on the continental shelf. As of 2014, they had five laurel leaf bifaces from underwater contexts (Stanford et al. 2014).

In 2000, a palaeo-reconstruction of the landscape of Block Island, Rhode Island was conducted using marine geophysics and sediment cores to determine the potential for archaeological sites. Though no archaeological remains were identified, the environment from 8000 to 10,000 cal BP, of an area that was once an island, was modelled. A total of 15-square miles were surveyed in three days using side scan sonar, a sub-bottom profiler, and the Datasonic seafloor imaging system. A rock dredge was used to collect seafloor samples. A gravity core was used to collect a core sample; however, coring attempts often failed due to rocky seafloor. The sub-bottom data were utilised to select specific locals for sediment cores (Coleman and McBride 2008, 205–10).

Westley (et al. 2011) modelled the landscape potential for archaeological sites on the northeast coast of Newfoundland, Canada. Based on the Danish model (Benjamin 2010), this project developed a seven-stage approach beginning with reconstructing sea level history. Second was mapping coastal evolution for detailed evaluation. Third was mapping buried landscapes using sub-bottom profiles. Fourth is sampling the seafloor using grab samplers and samples taken by an ROV or divers. They also used diver-operated hand-cores and cores taken using the sea-ice as the coring platform. Fifth was a creating 3D evolutionary model of the submerged landscapes using the sub-bottom data. Sixth was mapping the archaeological potential based on landscape attributes favoured by past humans to identify landscape settings with the highest potential for site preservation. Seventh, was archaeological testing using various methods based on water depth, including diver surveys, dredging, and grab samplers. They have conducted an intertidal survey of the study area of Back Harbour, and this survey identified a concentration of non-diagnostic lithic artefacts (Westley et al. 2011, 141). This project has incorporated the Canadian Maritime provinces including Newfoundland, Labrador, Prince Edward Island, and Nova Scotia conducting local and regional scale analyses including palaeo-landscape reconstructions and marine geophysical survey. They have more recently concluded that a few of the regions analysed have concrete evidence of human presence (Lacroix et al. 2014).

2.6. Conclusion

Underwater archaeology and high potential models often utilise a landscape approach and GIS technology. The examples presented in this chapter illustrate the methods and extent of research currently in both underwater archaeology and high potential modelling. The research is progressing quickly, and modelling needs to adhere to the theoretical frameworks from landscape approaches. As the capacity of GIS and computer technology continue to increase, high potential modelling will expand in extent and detail. Knowledge of underwater archaeology is expanding as a result of several edited volumes published in the 2010s.

Although the modelling aspect of this research is similar to other case studies, it is unique in the large areal extent, over 45,000 km^2 and in the refined scale of investigation, two-meter resolution. These attributes increase the computing requirement and increase the time required for model development and analysis. The theoretical frameworks described here provide clear methods and requirements to guide this research.

3

Introduction to the Pre–10,000 cal BP Palaeogeography of the northern NWC and Vicinity

3.1. Introduction

The Northwest Coast of North America is a narrow arc of land on the North Pacific Ocean stretching from the Copper River in the Gulf of Alaska southward to the Chetco River on the southern Oregon Coast (Suttles 1990b, 1). The coastline mainly runs parallel to the coastal mountain ranges, including the Chugach and Cascade ranges. The region is somewhat crescent-shaped, thin at its northern limit and expanding to a width of 320 km (200 miles) near the centre and then narrowing to about 160 km (100 miles) on the southern limit. The coastal plain is narrow and is characterised by two-vegetation area: [1] Sitka spruce and western hemlock in the north and [2] western hemlock and Douglas fir south of Johnstone Strait. Glaciers extend to the sea at various places (Suttles 1990a, 16, 20–21).

The region has a wet maritime climate (Carrara, Ager, and Baichtal 2007, 229). Winters are wet and mild. Summers are cool and wet. The climate largely stems from the adjacent Pacific Ocean with westerly winds and adjacent mountains. This causes moisture to fall as heavy rains on the western slopes of the outer ranges of mountains, while the leeward sides have reduced precipitation. Mean temperatures decrease from south to north and sea to shore (Suttles 1990a, 17–18). The average January high temperature in Ketchikan (the largest community in the study area) is three degrees Celsius (38 degrees Fahrenheit) with average low temperatures of negative two degrees Celsius (29 degrees Fahrenheit). The average July high temperature is 18 degrees Celsius (64 degrees Fahrenheit) with low temperatures of 11 degrees Celsius (51 degrees Fahrenheit). Ketchikan gets 30.35 cm of rain on average in January. It receives 349 cm of precipitation on average a year. The wettest month is October with 51.53 cm of precipitation (Intellicast 2018). The upwelling of colder water offshore keeps the coast cooler in the summer and nourishes plankton, the base of the food chain. The temperature of the sea varies with location and season. The ocean temperature in winter is six degrees Celsius (43 degrees Fahrenheit) and 15 degrees Celsius (59 degrees Fahrenheit) in summer (Suttles 1990a, 18). Salinity is variable in relation to resources. Around river mouths, brackish water may decrease the availability of shellfish. For most of the region, the tide is of a 'mixed' type, semidiurnal with two unequal high and low tides a day (Suttles 1990a, 19–20).

3.2. Southeast Alaska

Southeast Alaska extends from Icy Bay in the north to Dixon Entrance and the international border between the US and Canada in the south. The region is bordered by the Pacific Ocean on the west and northern British Columbia on the east. It is 845 km (525 miles) long and 193 km (120 miles) from west to east, composed of a narrow strip of mainland mountain ranges and over 10,000 islands. The offshore islands form the Alexander Archipelago rise out the sea to a maximum height of 1100 meters (3609 ft.). The largest island is Prince of Wales Island (POWI) (figure 3-1) (R. J. Carlson 2007; S. O. MacDonald and Cook 2009; O'Claire, Armstrong, and Carstensen 1992, 11).

There are 11 large rivers in southeast Alaska: Alsek, Chilkat, Taiya, Katzehin, Anteler, Taku, Whiting, Stikine, Unuk, Chickamin, and Salmon (figure 3-1). Most of these rivers are relatively short, with an average length of 20 km. Characteristically, they have steep-sided u-shaped valleys with narrow floodplains, rarely greater than 3.5 km wide, and the Stikine has the only major delta (J. A. Johnson, Andres, and Bissonette 2008, 3–5). These rivers are all on the mainland. There are no major river systems on the islands in the Alexander Archipelago.

Geologically, southeast Alaska has four main regions: The Alexander-Wrangellia, Gravina Belt, Taku, and Coastal Mountain Batholith. The Alexander Terrane originated as an island chain in the tropics as a spine of ancient volcanic rocks fringed by limestone reefs. The volcanic islands and the ones that were added as the island chain travelled northward through continental drift are called the Wrangellia Terrane. The Gravina Belt is the shed sediments from the surface wrinkle in the Alexander-Wrangellia formation. The Alexander-Wrangellia-Gravina terranes collided and fused with the North American plate. As stresses were relieved in the eastern area, rocks that had been forced deep into the earth rebounded forming the Coast Mountains. The areas adjacent to Revillagigedo and POWI were under less stress than other areas but were still up-thrust (R. Anderson 1991; Baichtal, Streveler, and Fifield 1997, 12; Brew et al. 1991).

The shorelines of southeast Alaska are indented with numerous embayments and coves with estuary environments (Lee 2007; Moss 1998, 2011). Much of the region was once glaciated and the landscape is characterised by fiords and drowned valleys adjacent to the continental shelf, which has been flooded by the post-Pleistocene sea level rise. The environment is dominated by water: ocean, rivers, bays, straits, and rainfall. The region is a temperate rainforest (figure 3-1) (O'Claire, Armstrong, and Carstensen 1992), and faunal and floral resources tend to concentrate at the interface between

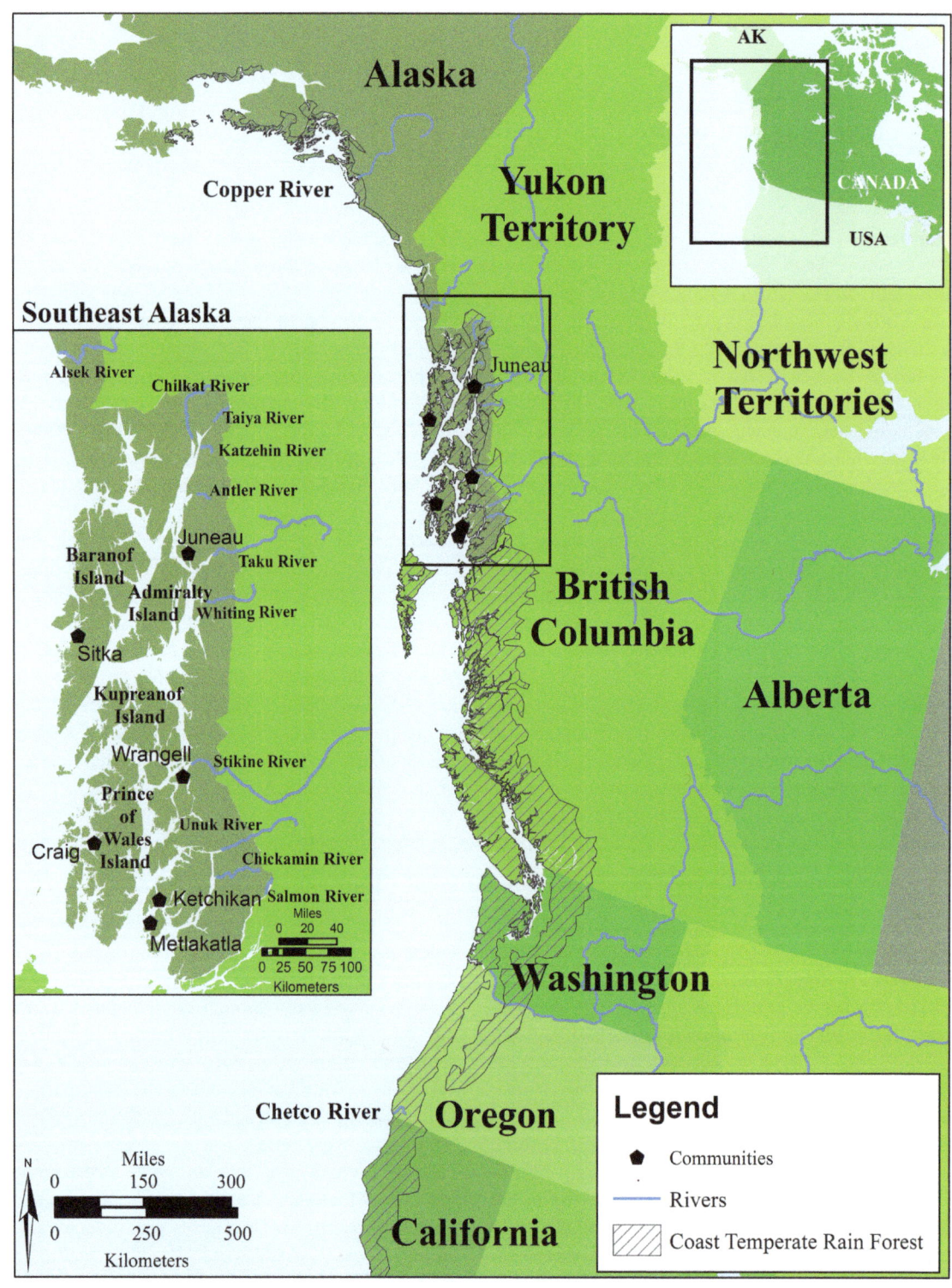

Figure 3-1. Map of the geographic region.

the land and sea in the intertidal and coastal zones (Moss 1998, 91).

3.3. Flora and Fauna

The lowlands are heavily forested with western hemlock (*Tsuga heterophyllai*), Sitka spruce (*Picea sitchensis*), and yellow and Western red cedar (*Chamaecyparis nootkatensis* and *Thuja plicata*) (S. O. MacDonald and Cook 2009, 26). Western red cedar and shore pine (*Pinus contorta*) are present in the southern half of the region. Yellow cedar extends from the middle of the Alexander Archipelago, Johnstone Strait, southward (Lyons and Merilees 1995, 76). Sitka spruce is drought intolerant and is mainly present along the coastline where there is more moisture. Western hemlock extends inland to the mountains (O'Claire, Armstrong, and Carstensen 1992, 19). The understory is dominated by huckleberry (*Vaccinium* spp.), devil's club (*Oplopanax horridus*), and alder (*Alnus* spp.) (S. O. MacDonald and Cook 2009, 26).

Muskeg vegetation is present over large areas of POWI in flat or gently sloping areas (Ager and Rosenbaum 2007, 4).

The sequence of plant colonisation since the LGM was originally investigated by Heusser (1955, 1960) and subsequently by Ager (Ager et al. 2010; Ager and Rosenbaum 2007) and others (Hansen and Engstrom 1996; Lacourse and Mathewes 2005; Warner 1984). The general sequence of plant colonisation began with tundra, approximately 15,000 – 12,500 cal BP, Pine is present around 12,500 cal BP, Spruce around 11,000 cal BP, and cedar around 5,000 cal BP (Fedje 2000). The timing for the appearance of cedar varies with earlier dates to the south and younger dates to the north indicating a northward migration of this species (Ager et al. 2010; Ager and Rosenbaum 2007; Reimchen and Byun 2005). This sequence of vegetation is approximate but corresponds for Haida Gwaii and southeast Alaska.

Southeast Alaska supports a rich mammalian fauna with some species found nowhere else in Alaska, such as the Pacific marten (*Martes caurina*), fisher (*Martes penanti*), four species of bats (*Lasionycteris noctivagans, Myotis* spp.), and five species of rodent (e.g. *Myodes gapperi, Peromyscus keeni*). Non-native species include brown rat (*Rattus norvegicus*), wapiti (*Cervus Canadensis*), and raccoon (*Procyon lotor*). The native mammalian fauna of southeast Alaska includes Sitka black-tailed deer (*Odocoileus hemionus sitkensis*), mountain goats (*Oreamnos americanus*), wolves (*Canis lupus*), black and brown bears (*Ursus americanus, U. arctos*), river otters (*Lontra Canadensis*), and mink (*Neovision vison*). Sea otters (*Enhydra lutris*) have been reintroduced and are now numerous along the outer coasts (S. O. MacDonald and Cook 2009, 29).

The region is also rich in marine mammals. These include Stellar sea lions (*Eumetopias jubatus*), harbour seals (*Phoca vitulina*), harbour and Dall's porpoises (*Phocoena phocoena* and *Phocoenoides dalli*), humpback whales (*Magaptera novaeangliae*), killer whales (*Orcinus orca*), and several species of beaked whales. In more recent years, California sea lions (*Zalophus californianus*) and elephant seals (*Mirounga angustirostris*) have become more numerous (S. O. MacDonald and Cook 2009, 29).

3.4. Glaciation

Since the Quaternary started 2.58 million years ago, there have been numerous glaciations. They are usually defined by marine isotope stages (MIS) based on oxygen

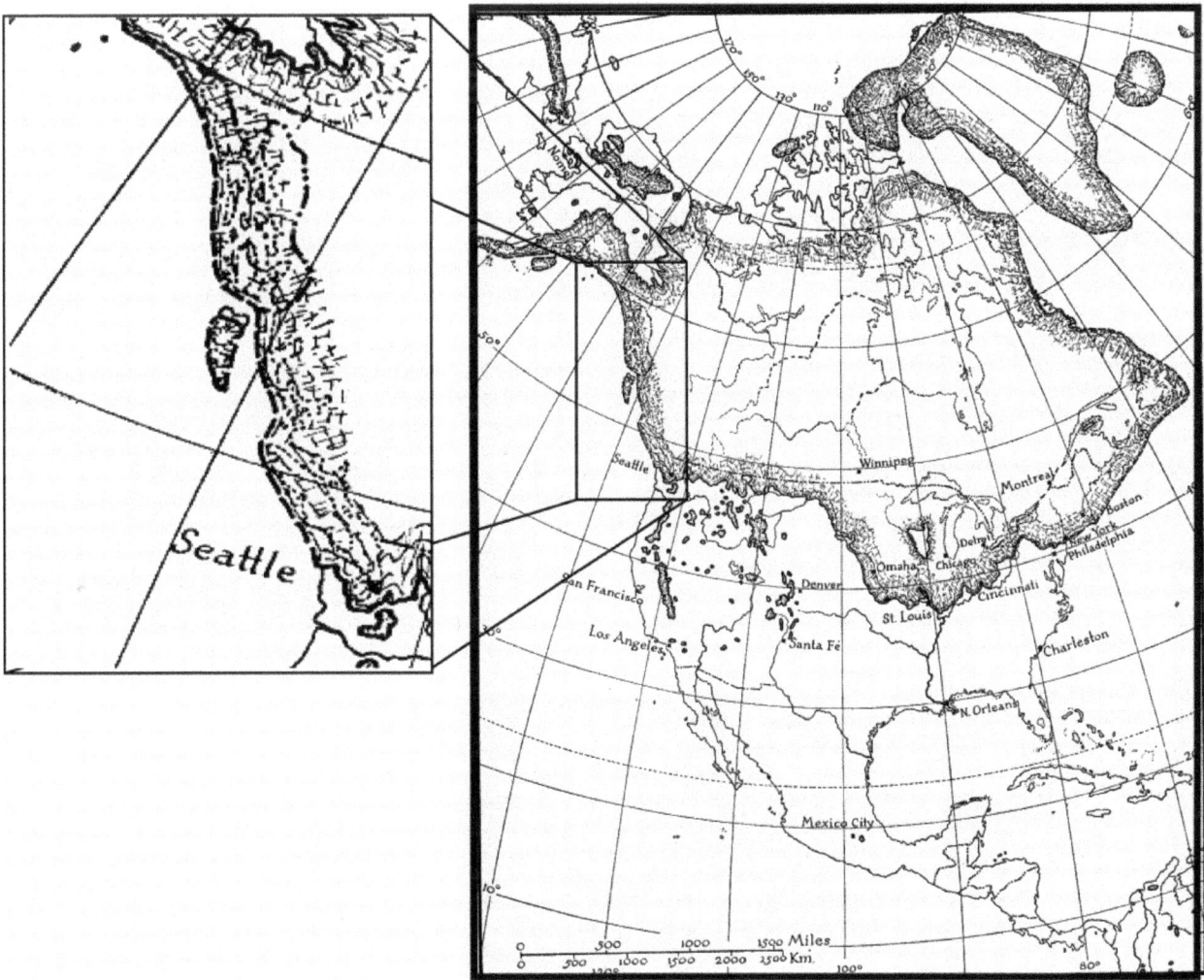

Figure 3-2. Map of Pleistocene glaciation from Antevs (1929, 651) with the NWC enlarged.

isotope data from deep-sea core samples. The most recent glaciation is from 110,000 to 10,000 years ago and is MIS 2-4. 'Wisconsinan' is the North American term for this glacial period. In British Columbia, the Wisconsinan glaciation is also called the Fraser glaciation (Blaise, Clague, and Mathewes 1990, 282). The maximum glaciation or last glacial maximum (LGM) for this period is thought to be approximately 21,000 BP, though it varied locally. Peltier and Fairbanks (2006, 3326) suggest that it might have been 26,000 years ago based on refinements to the Barbados sea level record. For Haida Gwaii, the glacial peak was between 19,000 and 17,000 cal BP; glacial retreat has been identified by 18,000 cal BP (15,000 ^{14}C years) (Blaise, Clague, and Mathewes 1990, 292; Lesnek et al. 2018, 4; Shugar et al. 2014). In southeast Alaska, the last glacial maximum was between 29,000 and 18,000 cal BP (25,000 and 15,000 ^{14}C years) (Clague, Mathewes, and Ager 2004, 86). The late Wisconsin glaciation is recorded from 30,700 to 15,700 cal BP (26,000 to 13,000 ^{14}C years) in southeast Alaska (Carrara, Ager, and Baichtal 2007, 231). A deglaciation corridor was open and a viable ecological pathway through southeast Alaska by 17,000 cal BP (Lesnek et al. 2018). Icy Straight was deglaciated by 13,500 cal BP (Mann and Streveler 2008).

In 1928, Ernst Antevs was asked to produce a volume summarising the Pleistocene glaciation (figure 3-2). He based his reconstruction on the assumption that during the largest glacial extent in British Columbia, the ice probably reached the open sea off the skerry-guard (outer islands or rock reef) (based on Dawson 1894, 286). The Queen Charlotte range was covered by an ice cap with glaciers descending the valleys and scouring out the fjords (based on (MacKenzie 1916, 117). Ice did not reach the northern shore of the island chain. Antevs differentiates the Admiralty and Vashon drifts on Vancouver Island. Unfortunately, Antevs was not able to determine which of the two glaciations (Admiralty or Vashon) was more extensive (Antevs 1929, 651).

For Alaska, Antevs used the USGS maps available at the time. Though generalising about the state, he realised that the glaciation was limited to the mountains and adjacent parts of the lowlands. 'The ever-present large and small fjords on the Pacific coast of Alaska were doubtless in part carved out and largely shaped by Quaternary glaciers, which joined to form a system of separate piedmont glaciers or a confluent piedmont glacier that probably covered the entire skerry-guard except the high ranges and summits' (Antevs 1929, 652). Antevs makes no specific comments about southeast Alaska. Capps (1931) suggested that large outlet glaciers of Wisconsinan age terminated on the outer continental shelf of southeast Alaska, similar to the modern coast of Antarctica (Mann 1986, 239). From Antevs' (1929) map (figure 3-2) of North America and his comments about both British Columbia and Alaska, it is clear that Antevs interpreted the Pleistocene glaciation as reaching the coast and encompassing the coastal region.

By 1986, the reconstruction of the glacier extent on the continental shelf based on ice flow models and onshore mapping suggests that the glaciers reached only the inner continental shelf in areas lying between major fjord entrances during the LGM (Mann 1986, Mann and Hamilton 1995: 460). Mann (1986, 260) suggests that Haida Gwaii was deglaciated by 16,000 cal BP and that the Alexander Archipelago may have been by that date. The glacial history of the Alexander Archipelago was likely complex during the late Wisconsin. Pleistocene age land mammals (Heaton and Grady 2003; Weckworth et al. 2011) and pollen cores indicate that both the Alexander Archipelago and Haida Gwaii had refugia or areas that remained unglaciated, during the LGM.

The Cordilleran ice sheet was a continental glacier that covered western North America from Washington north to the Yukon during the Wisconsinan glaciation. It is now known that Cordilleran glacial ice sheet flowed west to the Pacific from the interior mountains. It joined local glaciers from the high elevations in southeast Alaska as it flowed, though glaciers also formed on local mountains that did not join the continental glacier. These include glaciers on Admiralty, Baranof, Chichagof, and Prince of Wales islands. The large volume of ice was channelled into deep troughs, gouging out the present-day fjords, and providing the major outlet for the glaciers. Substantive areas along the outer coast were not glaciated; however, Dixon Entrance was one of the largest glacial outlets, as wide as 60 km (37.3 miles) (Carrara, Ager, and Baichtal 2007, 232). Other outlet glaciers included (figure 3-3): Icy Strait, Chatham Strait, Frederick Sound, and Clarence Strait (Mann 1986). These outlet glaciers scoured troughs visible on the bathymetry today. Coastal mountains blocked overflow of ice from the interior and limited glacier accumulation areas causing variation in ice extent on the continental shelf. 'Hence, throughout southeast Alaska during Late Wisconsin, glacier coverage on the continental shelf may have been locally variable and ice-free areas may have existed near the present coastline' (Mann 1986, 261). By 16,000 cal BP the continental shelf, fiords, and embayments of the Alexander Archipelago were mostly deglaciated. Some valley glaciers would have persisted for several centuries longer (Ager et al. 2010, 266; Mann and Hamilton 1995). Despite this detail, the western glacial extent is known poorly (Carrara, Ager, and Baichtal 2007, 232–33). The land is difficult to survey for glacial evidence due to the thick rainforest vegetation, the rainy weather, and lack of roads. Other evidence of glacial advances is submerged underwater but can be inferred in some areas based on bathymetry.

3.5. Sea Level Reconstructions

The archaeological implications of sea level change have been known for a long time, and early researchers used bathymetric maps to interpret seafloor geomorphology (Bailey and Flemming 2008, 2153). The sea level history for southeast Alaska is complicated by the glacial history. Due to the loading of the continental glacier on and adjacent to the region, a forebulge raised adjacent parts of the landscape. There is a presumed 'hinge' line/region somewhere in eastern Prince of Wales Island or further

Introduction to the Pre–10,000 cal BP Palaeogeography of the northern NWC and Vicinity

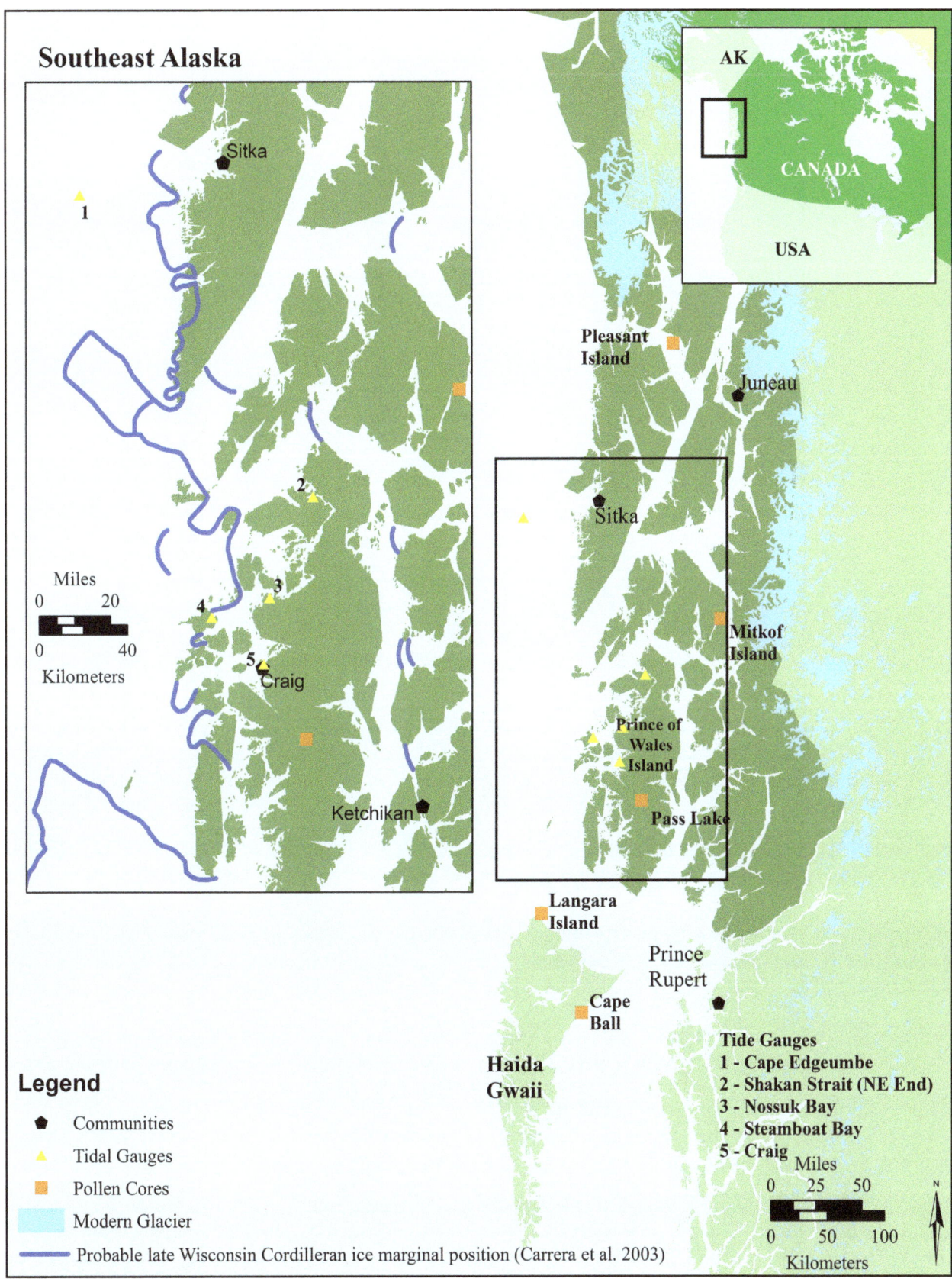

Figure 3-3. LGM glaciation and possible refugia.

east, where the land would have remained relatively stable compared to modern times. This is what has been proposed to the Hakai west sea level reconstruction (figure 3-4) where several pre-10,000 cal BP sites have recently been located (McLaren et al. 2014, 2015, 2018). Sea level curves are local phenomena and can vary based on geology, glacial loading, gravity, tectonic and other factors. A sea level reconstruction similar to those produced by

Uncovering Submerged Landscapes

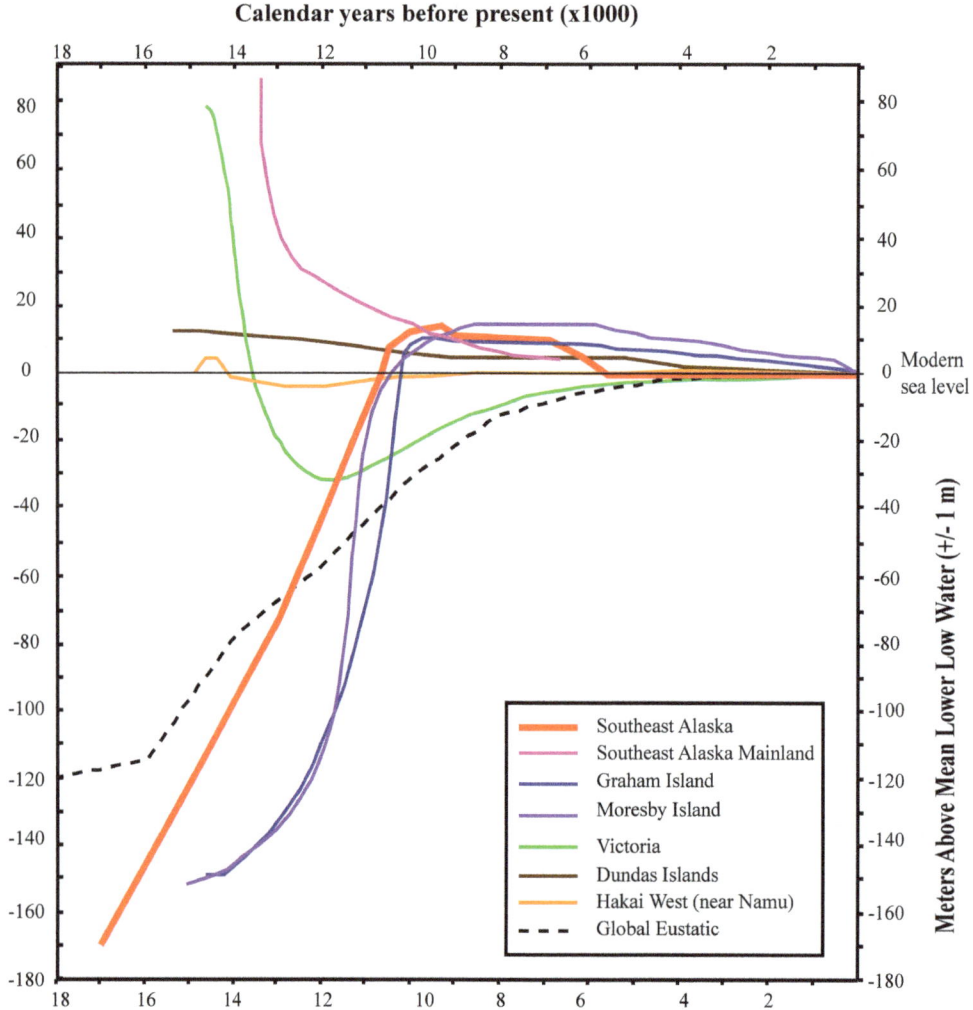

Figure 3-4. Sea level reconstructions for the NWC (Fedje et al. 2004; Shugar et al. 2014; Dixon and Monteleone 2014; Monteleone 2013).

Fedje, Mackie, and others for various locations along the Northwest Coast has been recreated here (figure 3-4) (Q. Mackie, Fedje, and McLaren 2018; Dixon and Monteleone 2014; R. J. Carlson and Baichtal 2015; Peltier and Fairbanks 2006).

During the LGM, global eustatic sea level was approximately 120 meters below modern levels (Bailey and Flemming 2008, 2153). Peltier and Fairbanks (2006) have extended the sea level reconstruction for Barbados back to 32,000 cal BP, based on the fossil coral record. Peltier and Fairbank's primary index points are from *Acropora palmate* because this species is ecologically restricted to five-meter below sea level. They then apply this sea level reconstruction to the globe using the ICE-5G (VM2) model (Peltier 2004). They have tested their model in various regions around the globe. Based on these data, they conclude that global sea level change since the maximum of 24,000 cal BP was approximately 120 meters.

There are three main processes at work in relation to sea level change: isostatic, eustatic, and tectonic. Isostasy is the internal processes of mass influencing topography (figure 3-5 A). Glacio-isostasy is the response to glacial and interglacial conditions changing the mass that is on the landscape (Ritter, Kochel, and Miller 2006, 23; Conrad 2013). The results of glaciation cause isostatic depression of the landscape and a forebulge where the land is elevated. Additionally, gravity lowers sea level near the glacier. Deglaciation causes isostatic uplift (figure 3-5 B). Eustatic changes are changes in the water level on a global scale (Ritter, Kochel, and Miller 2006, 37). Eustatic changes result from the changes in the volume and distribution of water in the ocean basins (Davidson-Arnott 2010, 27). Tectonic processes are caused by shifting tectonic plates or crustal movements, however, only the type of movement can be predicted, not the time or extent. Tectonic processes are sometimes dramatic and can cause sudden shifts in the landscape (Davidson-Arnott 2010, 23), including faulting events or earthquakes. The combination of these factors causes inundations (transgressions) or exposure (regression) of the land relative to the sea (Davidson-Arnott 2010, 19).

Southeast Alaska's sea level reconstruction is very complicated, partially because of the large geographic scale and the glaciers retreating to the east. Figure 3-5 C-F depicts a cross-section of southeast Alaska through Shakan

Introduction to the Pre–10,000 cal BP Palaeogeography of the northern NWC and Vicinity

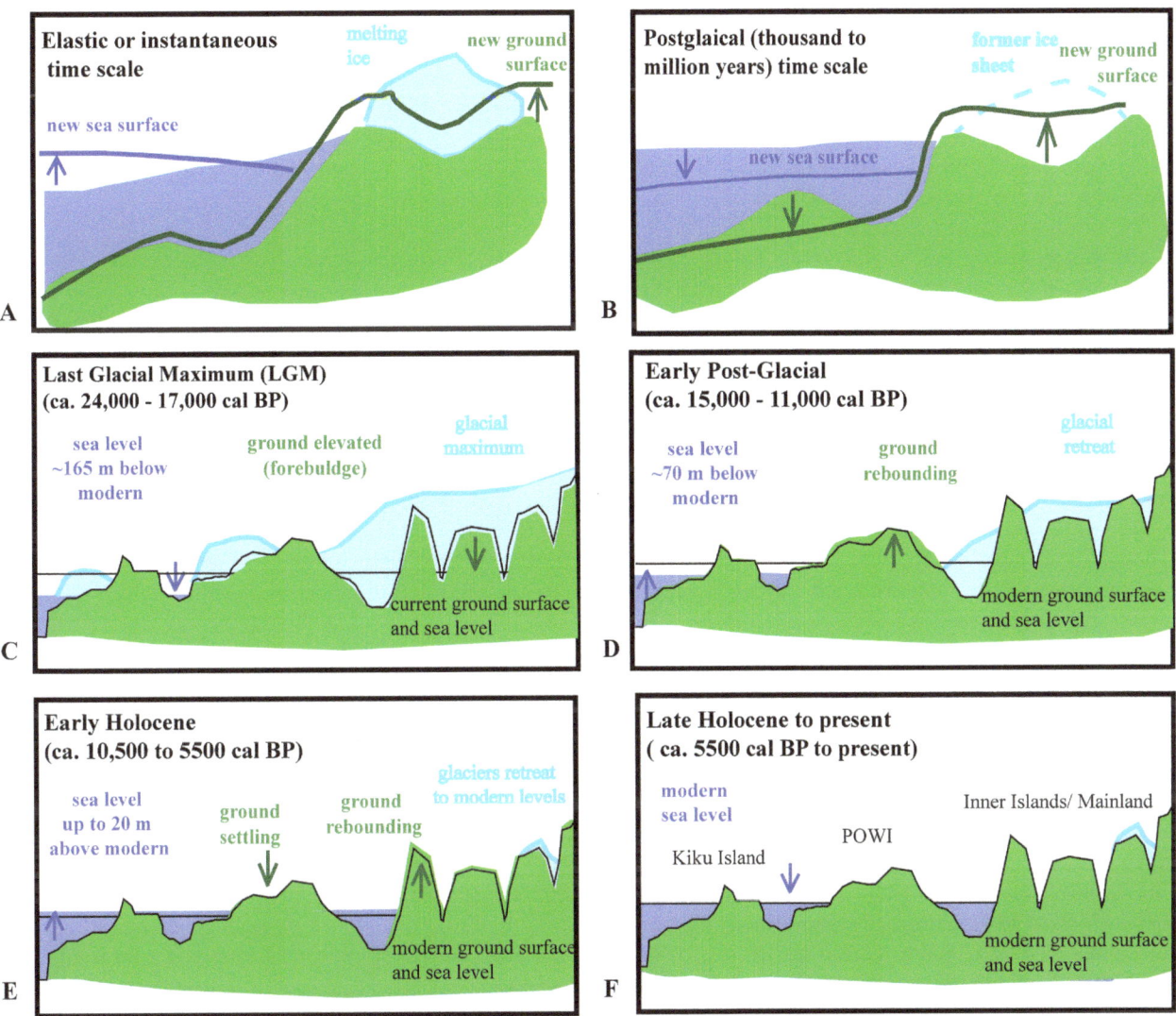

Figure 3-5. Changes in land-sea relation. A and B are models based on Conrad (2013) and C- F are reconstructions of a cross-section of southeast Alaska at Shakan Bay, similar to Fedje (et al. 2005, 24). A) Depicts the elastic or instantaneous movement of the land-sea relationship with glacial melt; here the land is rising due to isostatic rebound as the glaciers melt; the sea is rising due to eustatic sea level rise, with a small locally lower sea level near the land/glacier due to gravity. B) Depicts the post-glacial response; the sea level is lowering and the land where the glaciers once were rising due to isostatic rebound. The forebudge that is now under the sea as it collapses or lowers. C) Depicts the LGM approximately 24,000 to 17,000 cal BP when sea level was 165 meters lower than modern level. The reconstructed glacial extent is similar to 1600 meters ELA (Monteleone, Wickert, and Dixon 2017). D) Depicts an early post-glacial reconstruction from between approximately 15,000 to 11,000 cal BP where sea level was 70 meters below modern levels. The reconstructed glacial extent is similar to 2000 meters ELA. Here land that was deglaciated rises due to isostatic rebound. Sea level rises due to global eustatic sea level rise. E) Depicts the early Holocene from approximately 10,500 to 5500 cal BP where sea level is up to 20 meters above modern levels with modern glacial extent; the ground lowers or settles after the isostatic rebound to near modern levels. Sea levels continue to rise. Towards the mainland, the land is uplifted due to isostatic rebound. F) Depicts the late Holocene with modern sea level, glacial extent, and land surface; the major island of the cross section are labelled for reference (POWI = Prince of Wales Island). The thin black jagged line is the cross section through Shakan Bay associated with the modern sea level, or thin black horizontal line.

Bay. Figure 3-5 C shows the maximum glacial extent between approximately 17,000 and 24,000 cal BP with the lowest sea level for the region. Figure 3-5 D depicts the early post-glacial period from approximately 11,000 to 15,000 cal BP with lower sea levels and elevated land surfaces as the glaciers retreated. The reconstructed glacial extent is similar to 1600 m Equilibrium line altitude (ELA) based on an IceFlow model (Monteleone, Wickert, and Dixon 2017) where PET-408 would have been exposed throughout the glacial maximum (Heaton and Grady 2003). Here the land under the glacier is lowered due to an isostatic response from the weight of the ice; the sea level is lower due to the global eustatic sea level. Figure 3-5 D is the early Holocene from approximately 5500 to 10,500 cal BP when sea level was elevated above modern levels by up to 20 meters. Here there the modern glacial extent is depicted. Figure 3-5 E is the late Holocene with modern sea level, glacial extent and land surface since approximately 5500 cal BP. Modern glacial extent is a bit of a misnomer as, in southeast Alaska, the glaciers are presently thinning at a rate of up to 10 meters per year. Glacier Bay is also having isostatic rebound of up to 30 mm per year due to the collapse of the Glacier Bay Icefield. This is currently one of the fastest uplift rates in the world (Shugar et al. 2014, 176). For this model, 'modern' land surface is the one that is currently mapped and available.

3.5.1. Sea Level Reconstruction for the Outer Islands of the Alexander Archipelago

A preliminary sea level reconstruction for the outer islands of the Alexander Archipelago has been developed (figure 3-4 red line). Carlson (R. J. Carlson 2007; Baichtal and Carlson 2010; R. J. Carlson and Baichtal 2015; Williams and Dixon 2014) focused on the period when sea level was above modern (since 10,600 cal BP). She identifies three terrace levels where people could have lived. The Upper Terraces are between 16-21 meters above modern sea level and would have been occupied during the highest sea level (10,600 – 5,800 cal BP, 9,400 to 5,000 ^{14}C years). The middle terraces are between 8.5 and 13 meters. They were occupied from 5,800 to 2,000 cal BP. Finally, the lower or modern terraces are from six to seven meters and would have been occupied from 2,000 BP to present. This provides reliable dates for the part of the sea level record that is preserved above modern sea level.

There are currently only few points below modern sea level upon which to base the sea level reconstruction in the study region. The first is the -165 m point, which Baichtal refers to as a wave-cut terrace (Baichtal and Carlson 2010; Shugar et al. 2014, 180). In Sitka Sounds, west of Baranof Island has lacustrine diatoms underlying marine shells that date to 12,700 to 12,200 cal BP at a depth of -122 m; indicating the sea level was below -122 meters at this time (Shugar et al. 2014, 181). Barron et al. (2009) analysed and dated a sediment core (EW0408-11JC) from the Gulf of Esquibel, located immediately west of POWI. Their analysis indicates that when sea level was significantly lower, the Gulf of Esquibel was a freshwater lake. It was later flooded by the rising sea that ultimately formed the present-day Gulf. The core documents the Gulf as a freshwater lake between 14,200 to 12,800 cal BP that transitions to brackish water sometime between 12,800 and 11,100 cal BP. A radiocarbon determination run on shell by Barron (et al. 2009), adjusted for the marine reservoir effect, and calibrated to 11,324 – 11,773 cal BP also provides a minimum limiting date of approximately 11,500 cal BP. Based on contemporary bathymetry, the lowest topographic point along the basin's margins is approximate -70 m (Baichtal and Carlson 2010; Shugar et al. 2014). This is the probable point at which rising sea level breached the Esquibel basin.

There are geologic similarities between the outer Alexander Archipelago, which is of the Alexander-Wrangellia geologic terranes and Haida Gwaii, which is of the Wrangellia terranes (R. Anderson 1991; Shugar et al. 2014). Because of the similarity of the proposed sea level reconstruction for southeast Alaska (figure 3-4: red line) and that of Graham and Moresby Islands of Haida Gwaii (figure 3-4: blue and purple lines), it has been possible to construct a preliminary sea level reconstruction for the outer islands of the Alexander Archipelago. However, additional subsurface dating is needed to improve the reliability of this reconstruction.

3.5.2. Sea Level Reconstruction for the Inner Islands of the Alexander Archipelago

The inner islands have a very different sea level reconstruction than the outer islands. The possible glacial hinge area is still to be located within this region. As the glaciers retreated, the forebulge caused the elevation of the crust and subsequent collapse (figure 3-4: pink line) (Shugar et al. 2014). The shoreline rapidly moved landward. On Gravina Island, near Ketchikan, glacial marine shell deposits exist from 24 to 30 meters above sea level (asl). Within Behm Canal, there is New Eddystone Rock located 72 m asl that has been eroded by waves. In Portland Canal, glacial marine sediments are found from 103 to 152 m above mean sea level (Baichtal, Streveler, and Fifield 1997, 24). Although the sea level change for the inner island and mainland on southeast Alaska is complex, similarities can be drawn between the BC curves from Kitimat and Prince Rupert.

These data indicate that the region was largely covered by the Cordilleran ice sheet and its lobes from the east side of Prince of Wales Island to the mainland. This area was in a region of depression during the LGM that subsequently rose when the glacier retreated.

3.6. Last Glacial Maximum Refugia in Southeast Alaska

The proposed refugia locations for southeast Alaska are mainly on the continental shelf (figure 3-3). Heusser's (1960) work clearly demonstrated that parts of Haida Gwaii were not entirely covered by glaciers during the maximum of the Wisconsinan or Fraser Glaciation. In addition to pollen evidence for southeast Alaska, there are also faunal remains from before, after, and during the LGM (Heaton and Grady 2003; Carrara, Ager, and Baichtal 2007). Geologically, the evidence for refugia is based on the presence of local geomorphic features indicating that the land was either glaciated or unglaciated (such as U- and V-shaped valleys, glacial moraines, and deep gouges in the seafloor). In addition to the glacial evidence in southeast Alaska, researchers have also focused on contemporary and fossil faunal and floral evidence (Carrara, Ager, and Baichtal 2007).

3.6.1. Faunal Evidence Supporting Refugia

Heaton and Grady (2003, 19) started investigating discoveries of fossil vertebrates in the Alexander Archipelago in 1991. These well-preserved finds had been made in limestone caves by local spelunkers. They investigated 15 cave sites on Prince of Wales Island, surrounding islands, and the adjacent mainland. Of these 15 caves, eight date into the Pleistocene. The Devil's

Canopy Cave fauna provide insight into the fauna remains on POWI prior to the LGM (Heaton and Grady 2003, 22).

On Your Knees Cave (OYKC) contains some of the most significant cave deposits in the Alexander Archipelago. It is located on the northwest tip of POWI and has 10,300 cal BP old human remains and numerous cultural remains (Dixon 1999, 118). Heaton and Grady (2003) report the cave contains brown and black bear that predate and postdate the LGM. There is also ringed seal (*Phoca hispida*) remains that date to the glacial maximum. There were numerous late Wisconsin age bones recovered in the cave, many have puncture and gnaw marks from mammalian carnivores. Heaton and Grady (2003, 23–24) believe the cave served as a carnivore den both before and during the LGM but that many of the bones were reworked and deposited as a result of different depositional events.

In addition to the fossils, reported by Heaton and Grady (2003), several others have (Shafer et al. 2010; Shafer, Côté, and Coltman 2011; Sawyer et al. 2017; S. O. MacDonald and Cook 2009; Demboski, Stone, and Cook 1999; M. A. Fleming and Cook 2002) investigated phylogeographic patterns of animals to identify refugium. 'Genetic data of a mid-Pleistocene split of vertebrate taxa and geographic distribution of the mitochondrial lineages of stickleback, black bear, marten, and short-tailed weasel cumulatively suggest that a refugium existed on the continental shelf off the central coast of British Columbia' (Reimchen and Byun 2005, 93). Wigen (2005) suggests that the refugia were mainly on the continental shelf. The chestnut-backed chickadee (*Poecile rufescens*), Stellar jay, and northwest song sparrow (*Melospiza melodia*) showed increased genetic diversity on Haida Gwaii, which is a pattern consistent with a refugium. Other animals that are suggested to have survived in the refugia include black bear (*Ursus americanus*), garter snakes (*Thamnophis sirtalis*), water flea (*Daphnia pulex* complex), the lichen (*Cuvernularia hultenii*), long-tailed vole, Keen's mouse, and Mountain goats. They conclude there was 'near conclusive evidence' for refugia in the Alexander Archipelago and Haida Gwaii. Dawson caribou (*Rangifer tarandus dawsoni*) are likely a post-glacial colonist of Haida Gwaii. This caribou was a dwarf caribou adapted to the island. The last caribou on the island was shot in 1908 (Reimchen and Byun 2005, 82–83; Demboski, Stone, and Cook 1999).

Ermine (*Mustela ermine*) is a mammal that likely survived the LGM in the Alexander Archipelago refugia. There are three distinct lineages in southeast Alaska, based on mtDNA. Two have a wide distribution outside of southeast Alaska. However, there is an 'island' lineage found only on Prince of Wales, Suemez, and Heceta islands in southeast Alaska and Graham Island in Haida Gwaii (Carrara, Ager, and Baichtal 2007, 233–34). Fleming and Cook (2002) conclude that the 'island' lineage probably survived the LGM in coastal refugia.

Deermice (genus *Peromyscus*) have three distinct lineages that reflect divergence into at least three distinct ice-free regions throughout the LGM. This region is the extreme northwest range for deermice. The three refugia are [1] Beringia, [2] coastal near southeast Alaska and [3] southern continental. No locations were identified where these lineages are in contact in the Yukon. There is limited gene flow among island populations (Sawyer et al. 2017).

'Rather than a simple postglacial colonisation of a formerly inhospitable wasteland, there is a complex interaction of faunas – arctic and temperate, coastal and inland, Asian and North American – all interacting through a period of radical climatic change' (Heaton and Grady 2003, 45). The ring seal represents the arctic fauna. It breeds on landfast sea ice. The ring seal moved north when the southern limits of sea ice retreated northward. For refugium fauna, genetic studies of modern and fossil brown bear support the refugia theory (Heaton and Grady 2003, 46–47). Of the 108-mammal species or sub-species that inhabit Southeast Alaska, 27 are endemic to the region and 11 have ranges that mainly are confined to the region. These endemic mammals are more frequent on the outer islands (Baranof, Chichagof, Coronation, Forrester and Warren islands). This high degree of endemism and its focus on the outer islands 'has been interpreted as suggesting that refugia existed in some areas of the exposed continental shelf and outer islands of the Alexander Archipelago' (Carrara, Ager, and Baichtal 2007, 233).

3.6.2. Flora Evidence Supporting Refugia

Though Heusser conducted pioneering work with his pollen cores, none of his sites from Southeast Alaska south of Juneau was radiocarbon dated. They only provide a relative chronology of vegetation change. Others (Ager and Rosenbaum 2007; Ager et al. 2010; Hansen and Engstrom 1996; Warner 1984; Lacourse, Mathewes, and Fedje 2005) have continued these pollen cores in Southeast Alaska and Haida Gwaii. Figure 3-6 (locations on figure 3-3) combines these pollen cores into a chronology for the region. As areas were deglaciated or remained as refugia (Cape Ball), grasses, sedges and pine were predominate.

Similar to the faunal evidence, phylogeographic information is useful in providing evidence for the southeast Alaska refugia. Subalpine fir (*Abies lasciocarpa*), which is a mainland species, exists in small isolated stands on Dall, Prince of Wales, Heceta, and Kosciusko islands. It has been suggested that subalpine fir survived in refugia within the Alexander Archipelago during the LGM (Carrara, Ager, and Baichtal 2007, 234).

Shore pine (*Pinus contorta* var. *conorta*) was significant in the forest succession of the Pacific Northwest in early postglacial pollen records. 'The early arrival of shore pine (by 13,700 cal BP) and mountain hemlock (*Tsuga mertensiana*) at a site on Pleasant island in the Glacier Bay region has been interpreted as suggesting an expansion from refugia in the Alexander Archipelago' (Carrara, Ager, and Baichtal 2007, 234). There are also specific alleles in this variety of shore pine that indicate it was

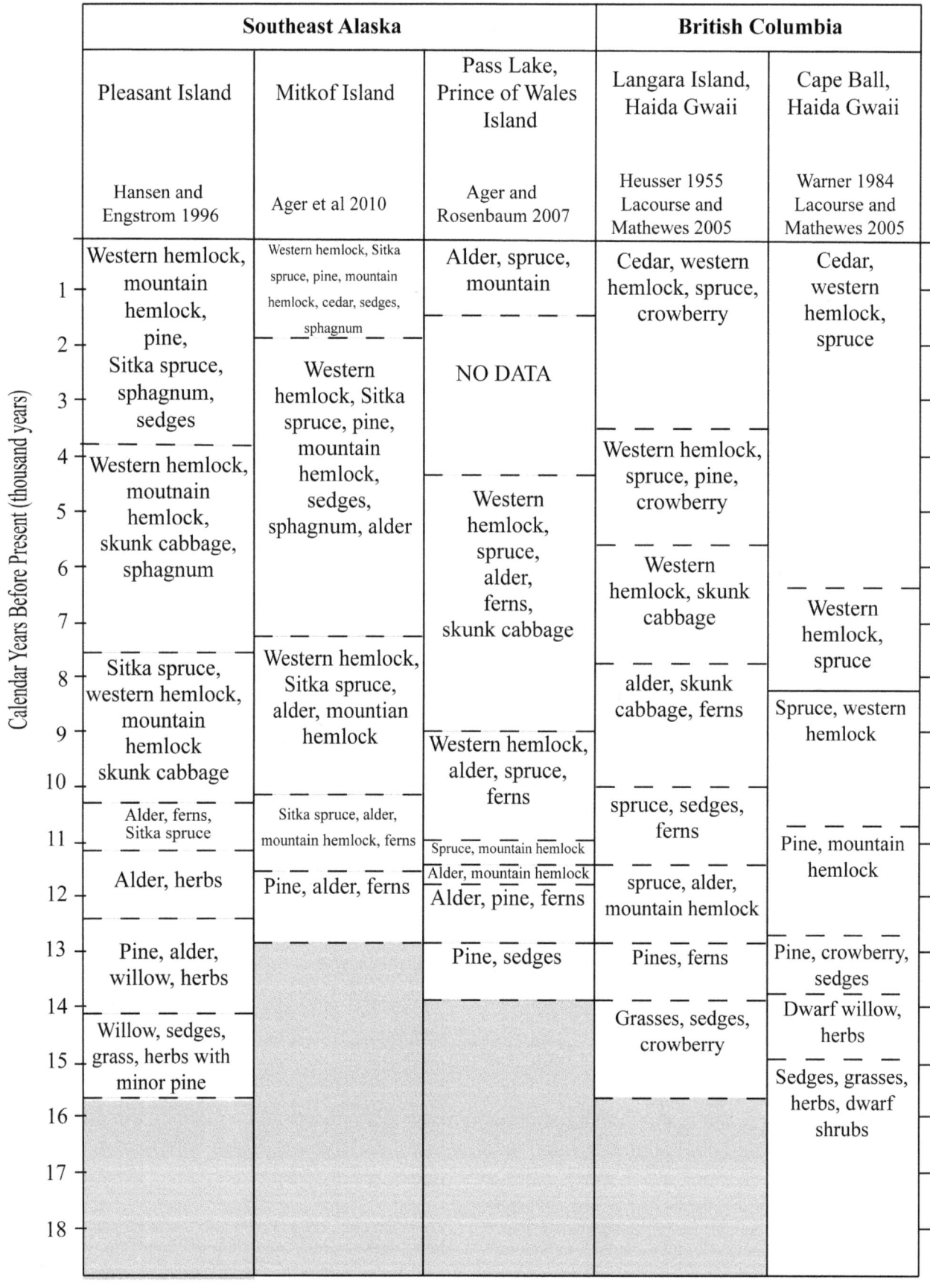

Figure 3-6. Flora pollen sequences (Ager et al. 2010; Ager and Rosenbaum 2007; Warner 1984; Lacourse and Mathewes 2005; Nixon et al. 2014; Dixon 2011b; Heusser 1955). See figure 3-3 for locations.

isolated for a considerable period of time. Because pine appears almost ubiquitously across the region soon after deglaciation, it suggests shore pine survived the LGM in coastal refugia. Shore pine then spread quickly across the previously glaciated landscape as the ice receded (Ager et al. 2010, 265–66).

Glacial refugia are well documented to have occurred on the outer islands of the Alexander Archipelago. A significant portion of the outer continental shelf of southeast Alaska would have been subaerial, unglaciated, and available for flora, fauna, and humans. Some areas of the continental shelf have undergone significant geomorphic changes as a result of post-Pleistocene sea level rise.

3.7. Geomorphology of Southeast Alaska

The dynamic nature of the ocean must be considered when investigating the shore and underwater environment. Southeast Alaska has substantial variation in the coastal processes. The region is characterised by rocky coasts with minimal beach areas. However, the complex coastline has sheltered areas where sites are likely to have been preserved as sea level rose (Benjamin 2010; A. Fischer 1995a). The shorelines of this region have changed significantly over the last 16,000 cal BP. The areas where coastal processes are relevant include tidal zones, the nearshore environment, and the continental shelf (figure 3-7). The west side of POWI, where this research focuses, does not have large rivers that would bury archaeological sites in deep sediment.

Generally, modern southeast Alaska has a meso-tidal variation (two to four meter) (Davidson-Arnott 2010), with unequal diurnal tides (Langdon 1977: 19-20). Southeast Alaska is dominated by tides with storm waves (Masselink and Hughes 2003, 4). The daily cycle is 24 hours, 50 minutes (Suttles 1990a, 18). The complex array of islands protects large areas of the coast from the full fetch of the Pacific Ocean and causes variability in the size, extent, and timing of the tides. The outer coast has a mean tidal range of 2.4 and 2.8 meters and a diurnal tidal range of 3.1 and 3.5 meters (table 3-1, figure 3-8). Currently, there is no accepted method to determine past tidal ranges. Consequently, modern tidal ranges, adjusted for the palaeo-coast, are assumed to be a good estimate (Davidson-Arnott 2010).

The nearshore is the portion of the coastal zone (figure 3-7) extending from the limit of significant sediment transport by waves to the low tide line. It is also defined as the shoreface (figure 3-7). This zone has the potential to be destructive to archaeological sites due to the movement of sediments. The nearshore environment is where most sediments move back and forth with the tides, the seasons, and storms. It can extend far off-shore during major storms and tsunamis (Davidson-Arnott 2010; Masselink and Hughes 2003). Sea level rose quickly in some areas, rapidly inundating the continental shelf and removing it from the nearshore movement of sediments (figure 3-4). This could have protected archaeological sites from long periods of erosion and extensive sediment loads and wave action. Additionally, there is no evidence for an environment favouring longshore drift in southeast Alaska.

The exact location of where the nearshore begins varies based on the size of waves within the region. Figure 3-8 shows the average wave height (m) from July 2002 to December 2008 from west of Baranof Island. The typical wave height is less than five meters, the period (T) is approximately seven seconds (Figure 3-8). Therefore, waves normally would affect the seafloor to a depth of approximately 38 m based on depth = 0.5λ and $\lambda = (g/2\pi) T^2$. The range of variability in the period is from 5.5

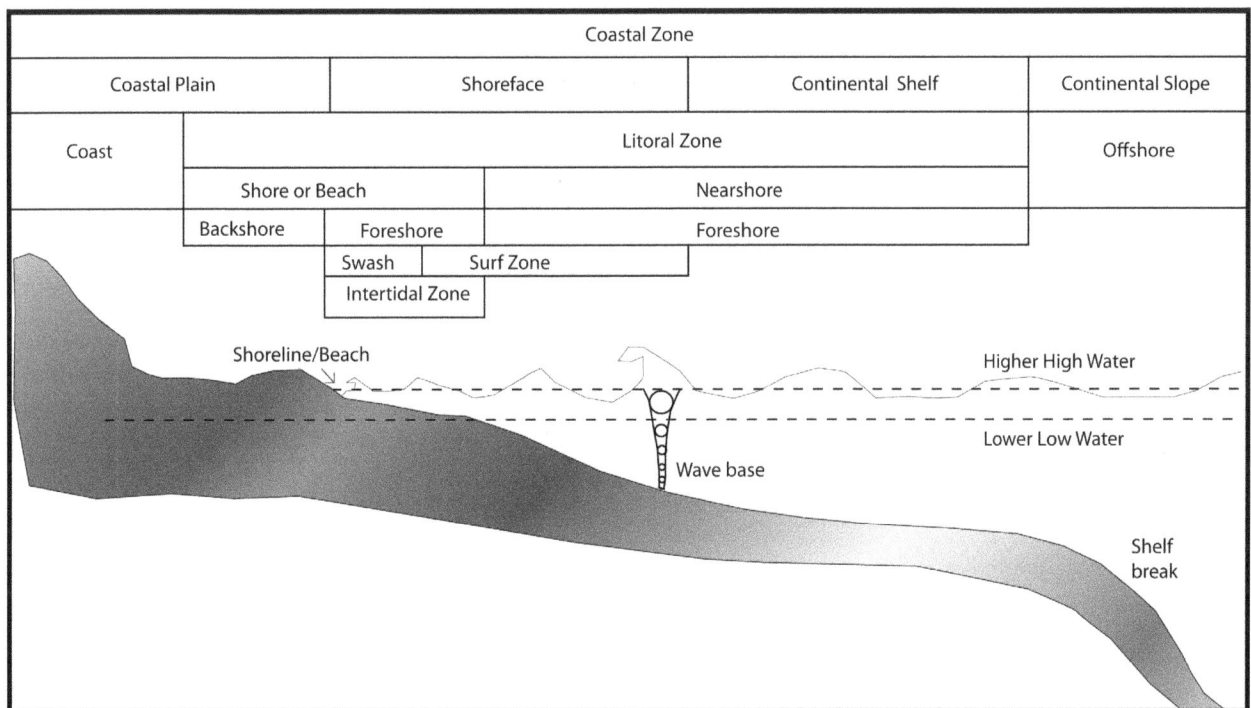

Figure 3-7. Coastal zone from the coastal plane to the continental shelf (Masselink and Hughes 2003, 1; Davidson-Arnott 2010, 12).

Table 3-1. Tidal range benchmarks around POWI in meters relative to mean lower low water (MLLW). See figure 3-3 for tidal gauge locations

	Station Number	Mean Higher-High Water (MHHW)	Mean High water (MHW)	Mean Tide level (MTL)	Mean Low Water (MLW)	Great Diurnal Range (GT)	Mean Range of Tide (MN)
Shakan Strait (NE End), AK	9450987	3.502	3.248	1.835	0.042	3.503	2.827
Nossuk Bay, AK	9450711	3.171	2.922	1.67	0.418	3.17	2.503
Steamboat Bay, AK	9450578	3.149	2.902	1.659	0.415	3.149	2.487
Craig, AK	9450551	3.101	2.842	1.63	0.418	3.093	2.424

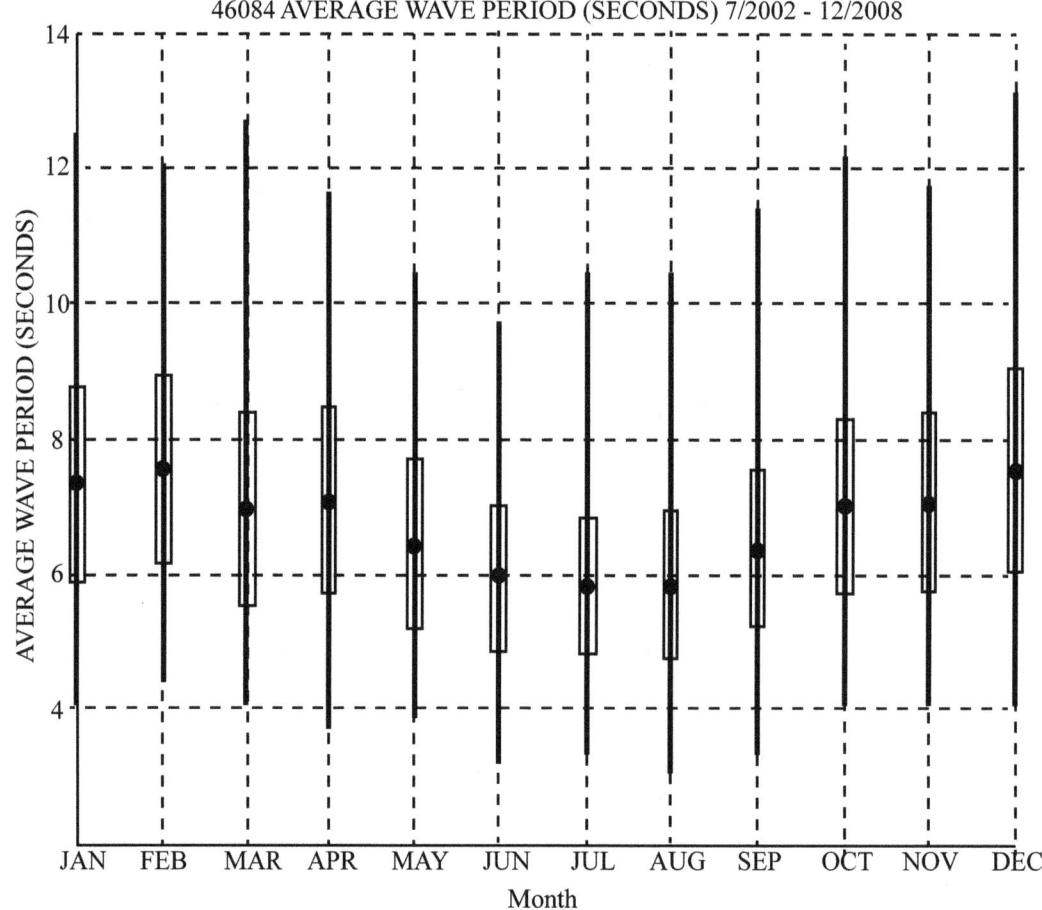

Figure 3-8. Average wave period from Cape Edgecumbe Buoy (46084) (NOAA, n.d.).

seconds to 8.5 seconds. This puts the range of transition from near shore to continental shelf between 23.6 and 56.4 meters below mean lower low water. The significant wave height of five meters is considered a near gale force wind (Wright, Colling, and Park 2005).

The continental shelf or offshore zone (figure 3-7) is the portion of the profile where there is no significant transport of sediment by wave action. This extends to the shelf margin, or where the shelf breaks to the deep ocean floor (Davidson-Arnott 2010; Masselink and Hughes 2003). This environment provides an area where archaeological sites could be preserved indefinitely, provided there is no change in the zone. A significant storm can push sediment onto a possible archaeological site or strip away any protective covering, depending on wave actions, tides, and where on the coast the transition is between the nearshore zone and continental shelf specifically is for that storm in relation to an archaeological site. There would have been many of these storms over the last 16,000 years.

3.8. Study Area – Shakan Bay

The Alexander Archipelago is a large area making it necessary to select a small subsection for survey (figure 3-9). This project focused on Shakan Bay in northwest POWI. Shakan Bay was selected because it is shallow and protected based on the preliminary analysis of the bathymetric data. It is south of On Your Knees Cave (OYKC) and in the ancient cultural interaction area

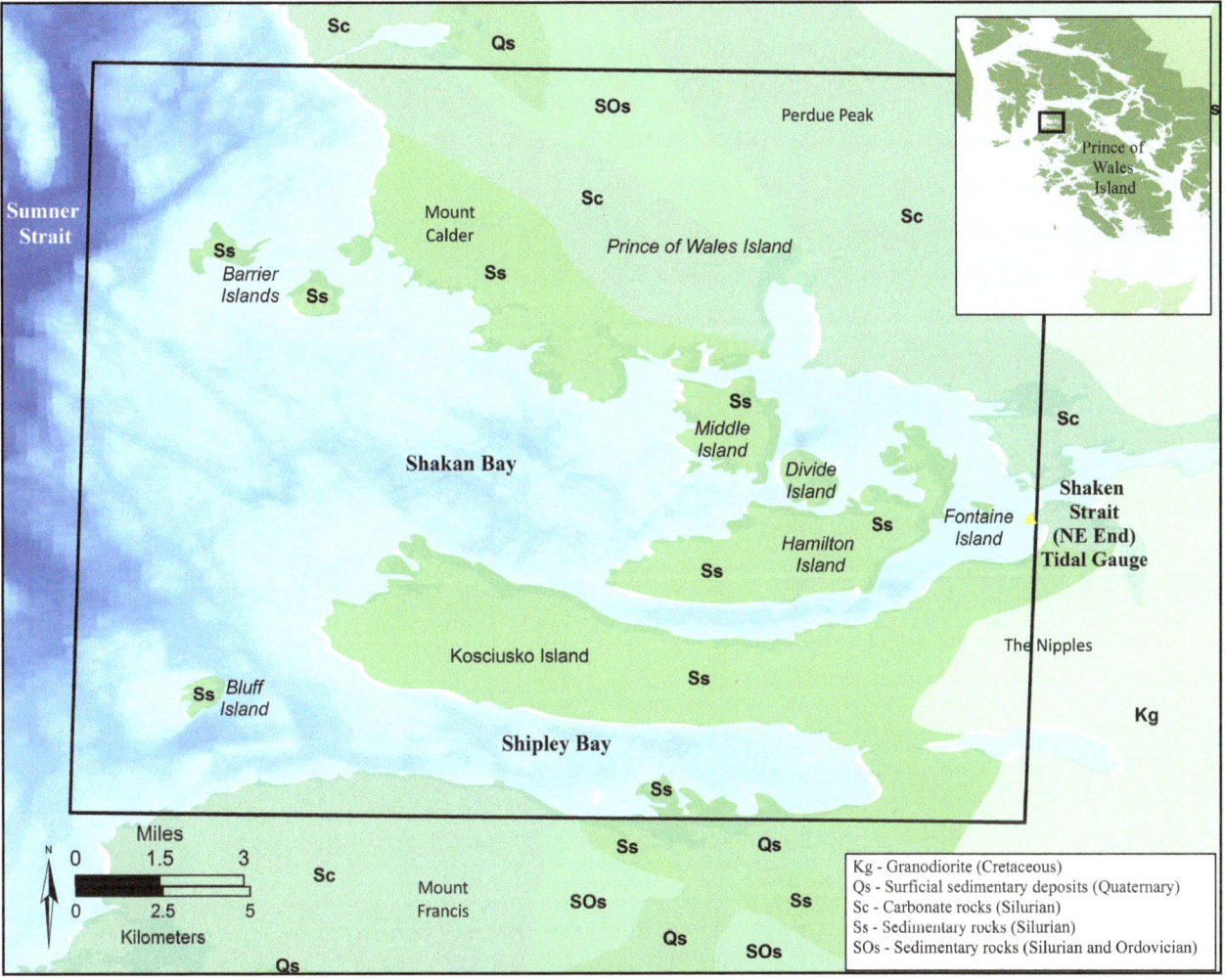

Figure 3-9. Map of Shakan and Shipley Bays depicting the geology (Gehrels and Berg 1992), the bathymetry, and locations named in text.

described by Carlson and Baichtal (2015; Baichtal and Carlson 2010).

There are two large bays within this area, Shakan Bay to the north and Shipley Bay to the south. To the west is Sumner Strait, which is very deep, over 200 meters from glacial scouring possibly from the LGM. The northern boundary of the region is POWI. The southern and eastern boundary is Kosciusko Island. Within the Shakan Bay, there is Hamilton, Middle, Divide, Fontaine, and Barrier islands (figure 3-9). The surrounding land includes several peaks and mountains, including Mount Calder, Perue Peak, and Mount Francis. Finally, Shakan Bay is connected to Shakan Strait where it wraps around Hamilton Island. The total study area is 319 km^2 (123 sq. mi).

The geology of the region was determined by Gehrels and Berg (1992) and is depicted in figure 3-9. The geology includes granodiorite, surficial sedimentary deposits, carbonate rocks, conglomerate, and other sedimentary rocks. These rocks and sediments are of the Ordovician, Silurian, Cretaceous, and Quaternary periods. The submerged surface geology, based on the NOAA map sediment indicators, is mainly mud, with some sand, and includes rocky and shell areas. The mud and sand would provide excellent environments to preserve archaeological sites.

The morphology of the region changes as sea level lowers. From zero to 25 meters depth, this region has two large bays. From 25 to 75 meters depth, the region had two intertidal estuaries. Landscape reconstruction suggests that these areas would have been highly productive for ancient people to collect, fish, and/or hunt. Below 100 meters, the region is mostly terrestrial with some rivers. In Shakan Bay, there is a deep valley-like trough through the centre that that has been interpreted to have been a river valley, lakes, or estuary when the bay was exposed as dry land. The adjacent estuarine environments would have provided sheltered areas where ancient archaeological sites could have survived sea level rise.

4

Archaeology and Ethnography of the northern NWC

4.1. Introduction

As a culture area, the northwest coast of the Americas (NWC) is defined by several cultural aspects, including distinctive woodworking technology; twined basketry decorated with false embroidery or overlay; basketry hats; emphasis on salmon harvesting; permanent villages or towns; and, social stratification with hereditary slavery (Suttles 1990b, 4). Geographically, the region is located along the Pacific Coast from the Copper River in the north to the Chetco River in the south (figure 3-1). The landward boundary varies by the researcher, but generally, it is within the coastal mountain ranges. The coast mountain ranges are the Chugach, Saint Elias, Coast Mountain, and Cascade Ranges. Though there is no question that the NWC is an ecological and cultural region, its exact boundaries are not precisely agreed upon (K. M. Ames and Maschner 1999; Drucker 1955; Kroeber 1939; Moss 2011, 9; Wissler 1914). The region has been subdivided many different ways, but most commonly into northern, central, and southern NWC (figure 4-1) (K. M. Ames and Maschner 1999; Suttles 1990b). The focus of this research is on the northern NWC.

There is substantial linguistic variability within the NWC, but culturally the groups are similar. At the time of Euromerican contact, there were over 40 languages belonging to a dozen language families (figure 4-3) (Suttles 1990b, 1). Many of the legends and myths (Boas 1982, 425), or oral traditions are shared or are common to various tribes. Other uniformities through the NWC include a lack of pottery and footwear, uses of plank houses, woodworking technology, and a heavy dependency on fish (especially salmon). There were permanent or winter villages or towns without large-scale agriculture (Suttles 1990b, 4), though some plants and molluscs were cultivated (Moss 2011). The NWC was a socially stratified society with inherited titles, commoners, and slaves, which is unusual among non-horticulturalists, if not unique. Slaves were captured from other tribes or born into their rank.

NWC archaeology centres on several key questions. First, the origins and land-use history of NWC people are important and is specifically relevant to this research. Second, there is significant archaeological variability in material culture from north to south along the NWC, specifically with respect to the presence and absence of early microblade technology, projectile point types, baskets, and other non-lithic artefacts. A third issue related to variability within this region is the development of the NWC cultures (Suttles 1990b). Finally, the cultural chronologies of the NWC are regionally variable. Different researchers have focused on different aspects of the archaeological and ethnographic records when developing different chronologies (figure 4-2).

4.2. Origins and Land-Use History of the NWC

The people living along the NWC have traditionally practised what is primarily a maritime-based economy. One of the questions that have been prevalent in the literature is whether the original settlers arrived maritime adapted (i.e. did they migrate to the region from other coastal regions) or were they terrestrially and riverine adapted (i.e. did they travel down rivers to the coast) and later developed a maritime adaptation. Maritime culture is an 'adaptation to the sea from which it [the culture] derives an essential part of the dietary and economic needs through the use of techniques and equipment particularly adapted to the exploitation of marine resources' (McCartney 1974, 158–59). Maritime economy (as defined for the circumpolar region) is 'a general culture-ecological pattern in which the economic base and general cultural orientation is partially or wholly dependent on coastal and maritime resources of the northern littoral' (Fitzhugh 1975, 343). One of the problems with early maritime adaptations was the perception that shell middens were created by the 'lowest class' because a shellfish gathering economy required only the most rudimentary cultural arrangements (Raab and Yatsko 2009, 16).

The early view was that people migrated down the rivers and slowly adapted to a maritime environment. Kroeber's (1939) model focused on adaptation rather than migration and diffusion (R. L. Carlson 1998, 24). Using the NWC and the Columbia-Fraser Plateau data, Kroeber divided the areas based on spatial continuity and environmental-adaptation criteria. He postulated that the NWC culture gradually developed from a river or river-mouth culture. It then developed into a beach culture. Eventually, and only in part, into a seagoing culture (Kroeber 1939, 28). He related the earliest stages of cultural development to the Columbia-Fraser Plateau and Intermountain Athapaskan. The intermediate stage, or beach culture, could be seen ethnographically in the Fraser River delta and islands. The open ocean or seagoing culture was related to the northern NWC groups including the Southern Tlingit, Haida, and Tsimishian (Kroeber 1939, 29–31).

Borden (1950, 1951) conducted archaeological excavations to test Kroeber's model. With data derived from the five sites that he excavated on or the near Fraser River delta, Borden developed a chronological sequence. The earliest is the Locarno Beach phase, which was a maritime adapted culture. He equates them with coastal Eskimo people. The next phase was the Marpole phase. Borden thought these people migrated down the river to the delta or river mouth.

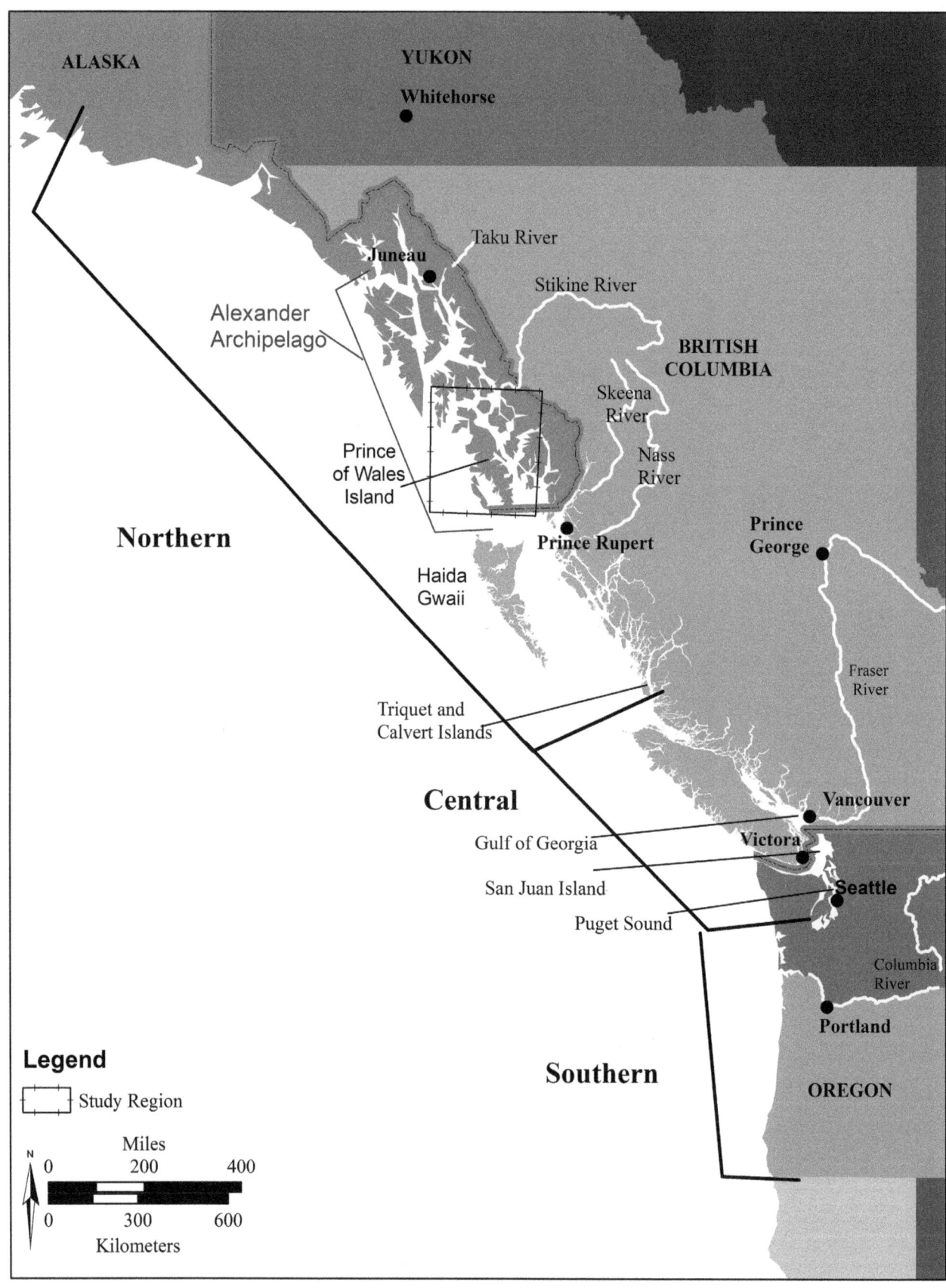

Figure 4-1. Northwest Coast including locations discussed in chapter.

King (1950) excavated at Cattle Point in the San Juan Islands. It was the first extensive excavation of a single shell midden on the NWC. He applied Kroeber's model of 'from land to sea' divided cultural development into three phases: 'Island', 'Developmental', and 'Maritime'. These transitions show an increase in bone and antler artefacts

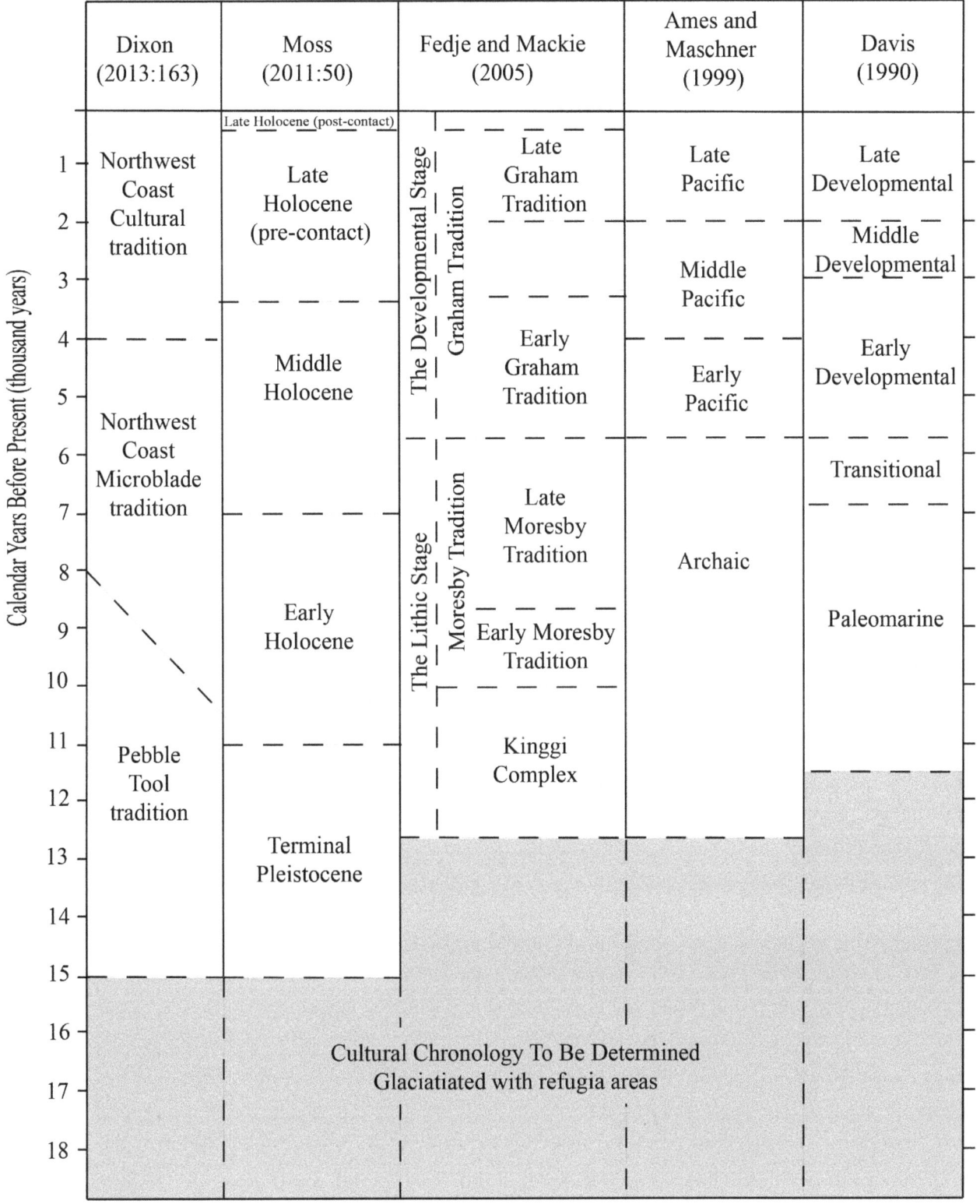

Figure 4-2. Published chronologies of the NWC (Dixon 2013a; Moss 2011; K. M. Ames and Maschner 1999; Fedje and Mackie 2005; Davis 1990).

adapted to exploit the sea (King 1950, 12). Cattle Point produced the first dark cultural layer with no molluscan remains under the shell midden (R. L. Carlson 1998, 25). King interpreted this as part of the 'Island' phase, but ideas that are more recent indicate the dark stain is a result of the dissolution of shell through percolating groundwater (Stein 1992) or possibly a house floor (R. L. Carlson 1993, 19–20).

Matson and Coupland (1995) suggested a model similar to Kroeber's. They propose that central NWC maritime cultures developed from Palaeo-Indian or Clovis hunters

Uncovering Submerged Landscapes

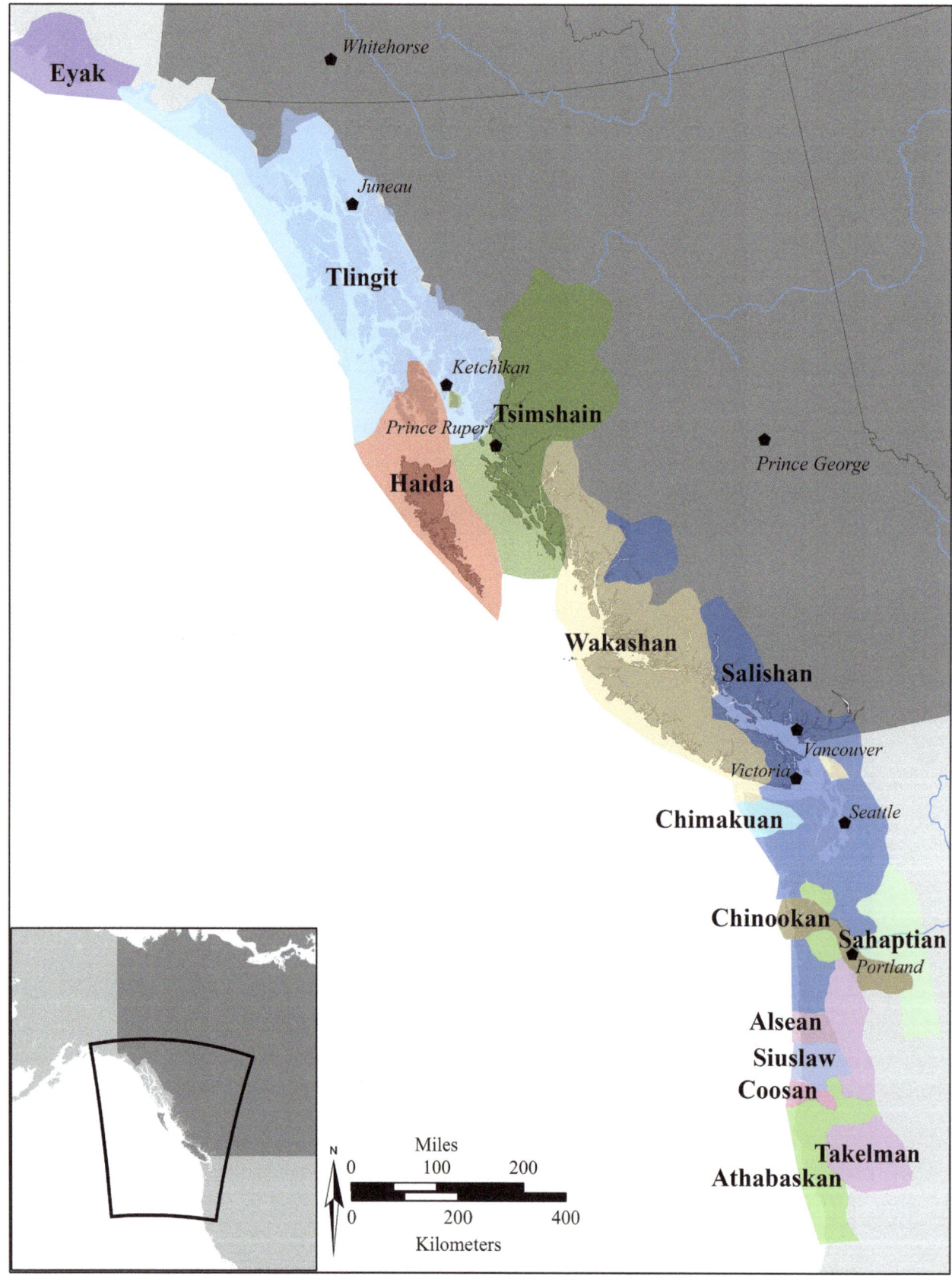

Figure 4-3. NWC language families based on Thompson and Kinkade (1990: 32).

from the interior of the continent. They advocate an initial migration from south to north for the central NWC and a north to south migration for the northern NWC.

By 1975, Borden had changed his conclusions. He proposed two separate initial migrations to the NWC. A northern group migrated down the NWC from north to south and a migration moved north from the Fraser River area (Borden 1975, 7). The Early Boreal tradition (or Northwest Coast Microblade tradition) migrated from Alaska and Yukon, seaward and down the coast. They already were adapted to a maritime economy and occupied the coast as far south as Namu. Borden (1975, 11) saw this group as wearing tailored skin clothing and having some means of watercraft, likely skin-cover frame boats from which they hunted and fished. The southern group used mainly unifacial tools (also called the Pebble Tool tradition by Carlson (1996)). Borden (1975: 84) uses the example of the Glenrose Cannery Site to demonstrate the early people were not able or engaged in hunting sea mammals. They were using intertidal resources by 8,300 cal BP (7430±340 ^{14}C years). According to Borden, the northern and southern groups mixed after 5800 cal BP (Borden 1975, 116). This seems to be a feasible option, though Borden's timing needs some refinements.

In contrast, Drucker (1955, 21) states that the earliest horizons in the Puget Sound areas 'represent a culture oriented towards the seas and the utilisation of marine resources, particularly the hunting of sea mammals'. Interior traits were subsequently adopted by this early maritime adapted tradition, as people adapted to the interior environment. Over time, the interior peoples lost their maritime economy. Drucker (1955) incorporates ethnographic information and describes each language group along the coast including their description of where they came from. These may not necessarily be the initial colonisation of the region but do constitute the earliest oral record. The Tsimshian tradition says that they came from 'a legendary place called Temlaxam ('Tuma-ham'), 'Prairie Town,' located somewhere far up the Skeena [River]' (Drucker 1955, 13). The Haida have been on Haida Gwaii 'since the creation of the world' (Drucker 1955, 12). The Tlingit used to live near the mouth of the Skeena River and have migrated northward (Drucker 1955, 11–12).

Early work had limitations that biased their initial assessment of the archaeological record because it was done prior to, or at the very earliest stages of, radiocarbon dating, or other absolute dating methods could not be applied to these sites. Dating was based on stratigraphy and relative positions. Secondly, the faunal remains were not usually analysed. In the 1970s, Carlson (1998) was told that there was no comparative collection of fish for the NWC. Finally, there is the preconception that maritime adaptations were a more recent development and were a 'lower class' resource (A. Cannon 1998; Maschner 1992; Raab and Yatsko 2009, 16).

Drucker's (1955) method of looking at the migration on a group-by-group basis is likely the most reliable and accurate method, though it does mean applying the oral history of modern ethnographic groups thousands of years into the past. Some groups likely came down rivers, like the Tsimshian and others likely came along the coast from either the north or the south (Moss 2004, 184). The archaeological evidence from the northern NWC has vastly increased and a maritime adapted migration has been gaining support for the initial colonisation of the NWC. The human remains discovered in On Your Knees Cave (OYKC; 49-PET-408) had an isotopic signature of an individual who was raised on a marine diet (Dixon 1999; Lindo et al. 2017). 'Whether they originated on the coast or in the interior is not clear, but they had knowledge of both' (Moss 2004, 184).

4.3. Cultural Chronologies

The cultural chronology for the NWC is still not clearly defined. Issues include the geographic variability and the lack of excavation on the northern NWC. There have been many attempts to unify the culture area into a single chronological sequence for the entire coast (figure 4-2). Several of these chronologies have been in use for over fifteen years (Moss 2004, 181) but all are lacking in at least one aspect, as discussed below. The chronology proposed by Fedje and Mackie (2005) incorporates many different aspects of the first three without issues with terminology, such as the use of pre-littoral. While the chronology presented by Dixon (2013a), though more recent, is very high level and does not include the detail see in Fedje and Mackie (2005). Moss (1998a; 2004, 2011, 50) uses a very simplistic but effective method of dividing the Holocene into early, middle, and late. This largely ignores the archaeology and uses environmental reasons for these divisions.

These chronologies were developed based on more complex location specific chronologies such as Fladmark (1982, 103). Here the NWC has been divided into five regions along the British Columbia coast. On the northern NWC, the Early Coastal Microblade tradition was represented by the Moresby tradition and a few other sites (Namu 1a and b, Skeena Crossing and the Paul Mason site). This complex has also been called the 'Early Boreal Tradition' by Borden (1975), the 'Early Coast Microblade Complex' by Fladmark (1975) and the 'Microblade tradition' by Carlson (1979). Fladmark eventually groups everything that is pre-5800 BP as the Lithic Stage.

The Lithic Stage extends from at least 11,000 to 5,800 cal BP (Fladmark 1982). It can be divided into [1] the Early Coastal Biface tradition, which is also called the Pebble Tool Tradition (R. L. Carlson 1990; Fedje and Mackie 2005; Dixon 2013a) or the Old Cordilleran tradition by Matson and Coupland (1995) and [2] the NWC microblade tradition (Fladmark 1982; Dixon 2013a; Fedje and Mackie 2005). The Early Coastal Biface tradition is characterised by leaf shaped points. On southern Haida Gwaii, karst caves dating to 12,500 cal BP (10,500 ^{14}C years) have the earliest evidence of this tradition (Fedje and Mackie 2005: 156). There are suggestions that these leaf-shaped points are derived from the Nenana tradition in Alaska or its Beringian antecedents (Carlson 1998, Dixon 1999, 2002, Fedje and Mackie 2005). The NWC microblade tradition is characterised by microblades, microblade cores, flake cores, pebble tools, and retouch flakes. Rarely are bifaces, abrasive stone, and ground or notched sinkers found in association with NWC microblade technology. Usually, the NWC microblade tradition is in non-shell midden contexts (Fedje and Mackie 2005, 156). The NWC microblade tradition is different from the American palaeo-arctic tradition and Denali complex found in other parts of Alaska indicating that the NWC people adopted or learned about the process of creating microblades from older and more formal traditions to the north (Dixon 2013a, 153). Dixon (2013a, 163) indicates that the Northwest Coast Microblade tradition started earlier in the north, 10,600 cal BP and later in the south, approximately 8,000 cal BP.

After the Lithic Stage is the Developmental Stage from 5,800 cal BP (5,000 C^{14} BP) to contact, it was named for the development towards the ethnographic cultural present (Fladmark 1982). The Developmental Stage was divided into Early, Middle, and Late. Many of these chronologies share a division at about 5,800 cal BP, but the divisions between the Early and Middle Developmental phases are different for each researcher. Davis' (1990) chronology is based partly on Fladmark's (1975). It incorporates the Developmental in Early, Middle, and Late Phases. Following Fladmark (1975), Davis' chronology also includes a Transitional Phase, which is seen in Lyman's use of the term 'pre-littoral' in a similar way that Davis uses 'paleomarine'. However, Lyman uses the term based on the exploitation of littoral resources in Oregon (Lyman 1991). Moss and Erlandson (1998b, 15) take issue with this term because 'pre' indicates that it is before the littoral period and is thus a contradiction in terms. The Palaeomarine phase name is derived from coastal-marine subsistence at the Hidden Falls, Ground Hog Bay 2, and Chuck Lake sites (Davis 1990, 197–98). Ames and Maschner (1999) divide the chronology into Archaic, Early Pacific, Middle Pacific, and Late Pacific. Their reason for using 'Pacific' was that other qualifying terms are unclear. Dixon (2013a, 163) places the transition to the Northwest Coast cultural tradition slightly later than the other chronologies at approximately 4,000 cal BP. His chronology focuses on the lithic technology utilised prior to 4,000 cal BP for defining the cultural traditions.

The Developmental Stage extends from 5,800 to 400 cal BP (5,000 to 250 ^{14}C years). It is divided into Early, Transitional, Middle, and Late. The Early phase dates from 5,800 to 3,800 cal BP (5,000 to 3,500 ^{14}C years) (Fedje and Mackie 2005, 156). It can be correlated with the Charles phase from the Gulf of Georgia that dates from 7,400 to 3,400 cal BP (6,500 to 3,200 ^{14}C years) (Easton 1993). The Early period is marked by the occurrence of microblades that are replaced by bipolar core technology, the beginning of ground stone and ground slate lithic technologies, bone and antler wedges, points, bone awls, and shell middens (Fedje and Mackie 2005, 156–57). The Transitional phase is suggested by Fladmark (1982) to be associated with the Locarno Beach phase on the south coast as between 3,800 to 2,600 cal BP (3,500 to 2,500 ^{14}C years). The Middle phase is associated with Marpole and dates from 2,600 to 2,300 BP (2,500 to 1,500 ^{14}C years). It is characterised by the traditional NWC cultural pattern including art, woodworking, plank houses, ornaments, exotic goods, and warfare. The Late phase is from 2,300 to 400 cal BP (1,500 to 250 ^{14}C years) and it is associated with the final development of ethnographic NWC cultures (Fedje and Mackie 2005, 157). All these chronologies recognise there is continuity throughout the NWC through time.

4.3.1. Variability in Site Locations

This research utilises the known archaeological site locations from the study region to develop variables for possible site locations in the past. This means that sites dating to the last 5,800 cal BP (5000 ^{14}C years) are combined for statistical analysis (Monteleone 2016). To test the hypothesis that there were no significant changes in site locations for the last 5800 cal BP, paired t-tests (alpha of 0.05) were conducted using the chronologies in figure 4-2, except for Dixon 2013. The null hypothesis for the ANOVA was that the means for the variables are the same or have no significant difference. The variables compared for each site location were slope; aspect; sinuosity of the nearby coast; distance to the nearest coast; distance to the nearest lake; distance to the nearest tributary; and, distance to the nearest site. Slope is the change in elevation over a specified distance, one-m, in degrees. Aspect is the cardinal orientation of the landform. Distance from a water body (lake, stream, or tributary junction) is the euclidean distance of a site to one of these features.

The null hypothesis for the ANOVA can be accepted for all variables except slope (table 4-1). A further series of paired-t-tests were used to analyse slope, using the same null hypothesis and confidence interval. Only three pairings exist where the mean slope is different between the cultural groupings (table 4-2). The first is the late periods (which is the same for all four cultural chronologies) and Fedje and Mackie's (2005) early period with a significance value of 0.03. This pair is at each end of the temporal spectrum and thus some variability is acceptable. The second pair is between Moss's (1998) Middle Period and Ames and Maschner's (1999) Middle Period with a significance value of 0.03. This is possible because Moss's period extends 2000 years longer than Ames and Maschner's Middle Period. The final pair in which the null hypothesis was rejected is Fedje and Mackie's (2005) Early Period and Davis' (1990) Transitional Period where the significance value is 0.04. Here Davis' (1990) period extended beyond the other chronologies by over a thousand years to prior to sea level stabilisation. Given the demonstrated consistency in site locations through time after sea level stabilised around 5800 cal BP, this research assumes that there was minimal variability in site location selection further back in time. These site location statistics were used as a basis for the predictive model discussed in the next chapter.

4.4. Geographic Variability

Ethnographically and archaeologically, there is significant variation from north to south within the NWC region.

Table 4-1. Results of grouped ANOVA for study region; the null hypothesis can be accepted, or that the chronologies are statistically the same except for slope (in bold) as it has a p-value of less than alpha (Monteleone 2016)

		df	Mean Square	F	Sig.
Slope	Between Groups	8	397.71	2.32	0.02
	Within Groups	647	171.22		
	Total	655			
Aspect	Between Groups	8	14534.61	1.26	0.26
	Within Groups	647	11510.72		
	Total	655			
Sinuosity	Between Groups	8	2.85	0.31	0.96
	Within Groups	647	9.14		
	Total	655			
Distance to nearest Coast	Between Groups	8	363708.29	0.14	1.00
	Within Groups	647	2596802.33		
	Total	655			
Distance to nearest Lake	Between Groups	8	433800.60	1.10	0.36
	Within Groups	647	392862.47		
	Total	655			
Distance to nearest Site	Between Groups	8	330406.93	0.42	0.91
	Within Groups	566	778910.03		
	Total	574			
Distance to nearest Tributary	Between Groups	8	413809.57	0.59	0.79
	Within Groups	566	707082.86		
	Total	574			

Table 4-2. Results of paired t-tests for slope where the null hypothesis can be rejected (Monteleone 2016)

		N	Correlation	Sig.
Pair 7	Slope of All Late & Slope of Fedje Early	30	0.39	0.03
Pair 10	Slope of Moss Middle & Slope of Ames Middle	101	0.21	0.03
Pair 34	Slope of Davis Transitional & Slope Fedje Early	24	-0.42	0.04

This variability has led to differences in subsistence, social organisation, and technology (K. M. Ames and Maschner 1999). There is also a difference in the number of archaeological surveys and excavations. As with many regions, areas adjacent to large contemporary populations have more archaeological surveys, such as around Vancouver and Victoria, BC, but the area around Puget Sound and the Fraser River have been the focus of more archaeological work than the northern region (R. L. Carlson 1990, 113). Archaeological surveys for logging are the exception in the northern region. These are the most common archaeological survey. These survey location biases are counterbalanced by research projects on Haida Gwaii, specifically in Gwaii Haanas National Park and Reserve (Fedje and Mathewes 2005) and though recent funding from the Hakai Institute has expanded work in remote areas along the BC coast (Letham et al. 2015; McLaren et al. 2014; Q. Mackie, Fedje, and McLaren 2018).

One of the earliest distinctions between north and south was the initial tool technologies. On the north coast, microblades were present very early (before 8800 cal BP) (K. M. Ames and Maschner 1999, 165; Dixon 2013a). The southern areas had stemmed points and crescents similar to the continental interior (R. L. Carlson 1990, 61). By Middle Pacific Period (3500 to 2000 BP), there were north-south distinctions in art styles, including whalebone clubs, cranial deformation, and labret wears (R. L. Carlson 2000, 170). The Middle Pacific Period also includes common practices of house and town organisation.

Tlingit houses were rectangular with a low-pitched gable roof. The entry was an oval in the front of the house at the usual height of the winter snow. The four main posts were often carved and painted (de Laguna 1990, 207). Haida houses are made of red cedar timbers and planks. There are two basic forms: a seven-beam roof and a four-beam roof style. The larger houses included an excavated floor to increase the size of the house. In front of Haida houses are carved totem poles occasionally positioned to provide the entryway of the house through a hole in the stomach or mouth of something on the pole (Blackman 1990, 243). Bella Coola has two and six beam options with a gable roof. Some of the entryways were part of totem poles while others were simple rectangular openings. Wakashan has one and two beam house options with a gable roof. On the southern coast, there are shed-roof and gambrel-roof houses in addition to the gable roof style (Suttles 1990b, 6–7). The Chinookans of the lower Columbia River had oblong, gabled-roof, upright-cedar-plank houses. Some of these were single dwelling (Silverstein 1990: 537), while the northern house included an extended family, plus slaves, of 40 to 50 people (de Laguna 1990, 207). In terms of underwater archaeology, such substantial structures should be detectable by remote sensing due to subsurface disturbances.

Statistical analysis (Monteleone 2016) indicates that the social structure did not affect site selection in the northern NWC, as there was no significant difference in site selection in 5800 years. The social structure reviewed was based on ethnographic records and it is difficult to determine how far into the past each of the social variables extends. The review of these social structures provides a comprehensive ethnographic review in the hopes of gleaning information about the distant past.

4.5. Northern NWC Ethnographic Groups

Three ethnographic groups are considered part of southeast Alaska and are within the study region (figure 4-4). The Tsimshian are mainly around the mouth of the Nass River on the mainland, but also have a reservation (the only one in Alaska) called Metlakatla on Annette Island south of Ketchikan Alaska. The Haida moved north into southeast Alaska at some unknown point in the past and occupy the southern end of southeast Alaska, including areas of POWI. The Tlingit have been in southeast Alaska longer than the Haida, but for an unknown amount of time. They are present throughout the region, but mainly north of the Haida. The Eyak are also in southeast Alaska but are north of the study region. The Tsimshian are east and south of the study region. Lindo and colleagues (2017) used DNA recovered from Shuká Kàa, the 10,300 cal BP human remains from OYKC, to demonstrate genetic continuity for the last 10,300 cal BP on the NWC. Additionally, the NWC population was on a different ancestral line compared with other early Holocene/ late Pleistocene individuals (Anzik and Kennewick Man).

4.5.1. Tsimshian

The Tsimshian are a group based on language and common cultural traits (figure 4-4). They live in northwest BC along the Nass and Skeena rivers and at the reservation at Metlakatla, AK. There are four major divisions: the Nishja on the Nass River, the Gitksan on the Upper Skeena, the coast Tsimshian on the lower Skeena and coast, and the southern Tsimshian on the coast and islands to the south (Halpin and Seguin 1990, 267). Coast Tsimshian and Nass-Gitksan form a language family. There may also be a third language or simply a separate dialect of southern Tsimshian (Thompson and Kinkade 1990, 31–33).

Each local group traditionally occupied a single winter village, moving in the spring and summer to fishing villages or camps. The winter villages were usually located near available freshwater, a variety of plant and animal food resources, sheltered from strong winter winds, and preferably in a position providing defence against attack. Most activities required transportation by canoe, as it is time-consuming to walk even short distances in the rainforest without cleared trails (Halpin and Seguin 1990, 269–71).

Tsimshian seasonal rounds are based on coast and southern Tsimshian historical accounts. When the river ice breaks up in the spring, the main activity was eulachon fishing on the Nass River. The fish was dried or processed into an eulachon oil (or grease). Their monopoly on the 'grease trade' made them very wealthy. May was for gathering and drying seaweed, usually for about a month. Herring spawn was caught in large quantities on grass, kelp or branches suspended in the water and dried for later consumption. Men would fish offshore for halibut. Women would cut them into filets for drying. Red cedar bark was collected for winter weaving. Hemlock, spruce, and lodgepole pine cambium (the layer inside the bark) was eaten contribution to a large number of culturally modified trees (CMTs). June was when sea gull eggs and oysters were gathered. Abalone was taken at the lowest tides of the spring. In the early summer, salmon began to enter the streams and the Tsimshian move to salmon fishing camps, which were controlled by the corporate matrilineage group (or house) and managed by the chief. Throughout the summer, women were active gathering berries from house territories and then drying them or preserving them in grease. Early autumn was an intensive salmon harvesting period. Chum salmon, which were ideal for preservation because of low-fat content, began to move into the rivers in the fall. Each house controlled several different fishing stations and had access to all five salmon species to provide insurance again famine due to a poor salmon run of any particular species (Halpin and Seguin 1990).

After the salmon were stored, hunting game became the major focus. Regularly hunted game included deer, elk, seal, sea lions, sea otter, mountain goats, mountain sheep, bear, porcupine, racoons, eagles, marmots, caribou, moose, mountain lion, hares, lynx, swans, geese, ducks, and other waterfowl. In the winter months, people remained mainly in the village. Shellfish were gathered including cockles, several varieties of clams, and mussels. Women wove, and men carved during the winter, which was also an important time for rituals and ceremonial events including potlatches (Halpin and Seguin 1990, 269–71). The winter villages, fishing camps (summer), and seaweed collecting camps (spring) should be visible archaeologically if they were occupied for several seasons and/or for the entire season.

Archaeology and Ethnography of the northern NWC

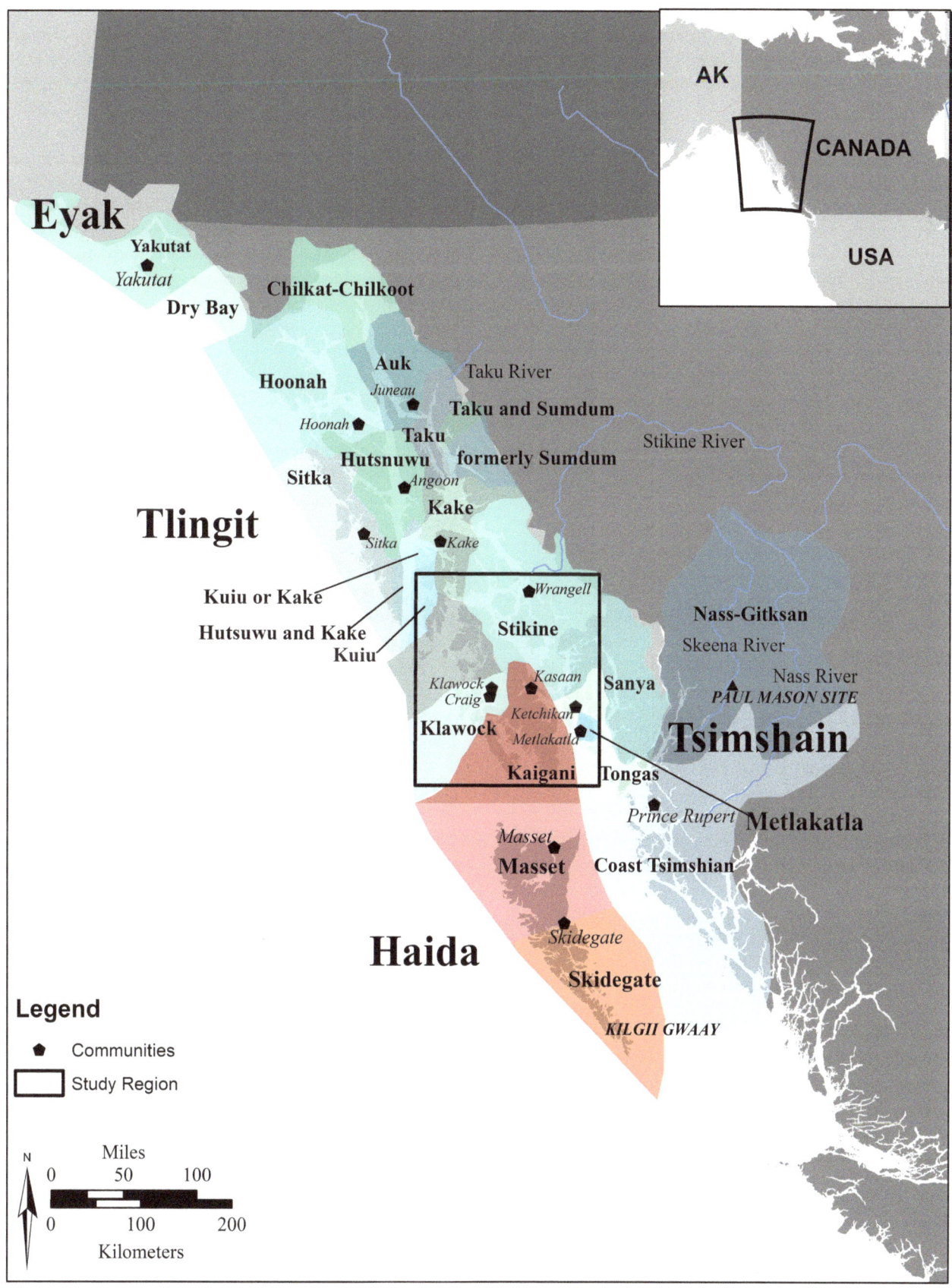

Figure 4-4. Northern NWC ethnographic groups based on Thompson and Kinkade (1990: 32).

Winter houses were constructed of massive red cedar timbers. Houses averaged 15 by 17 meters (50 by 55 feet). Thick upright planks on the four sides were grooved to fit horizontal planks, just outside the corner posts. Some people took their house planks to their spring and summer camps. The doors were either rectangular or oval and were at the gable end facing the beach. Some doors were cut into totem poles at the front of the house. The house names were inherited like crests. There were excavated pits for the central fireplace in the centre of the house (Halpin and Seguin 1990, 271).

Basketry had two main techniques that were important to the Tsimshian, plaiting (which is a type of checker pattern), and twining. Men made woodworking objects such as storage boxes, chests, the northern type of canoe, woodworking tools, and fishing and hunting gear (Halpin and Seguin 1990, 273–74).

Tsimshian have a four-clan structure consisting of Killer Whale, Wolf, Eagle, and Raven with a few variations (such as the Gitksan having Fireweed in place of Killer Whale). The basic social unit of the Tsimshian society was a corporate matrilineage called a 'house'. The 'house' owned fishing, hunting, and gathering territories and localities that were controlled by the house chief. Crests are images and symbols of privileges that were owned by a house and ceremonially displayed. Crests were represented on architectural features (totem poles, house entrance poles, house posts, house front painting, beams, rafters, and ceremonial entrances), costume features (robes and headdresses), feast dishes, and ladles. They were used during the potlatches. The village chief was the highest-ranking house chief within the village, but the role of 'chief' only was recognised in the coast and southern Tsimshian (Halpin and Seguin 1990, 274–76).

The first European contact was likely at the village of Kitkatla in 1787 by Captain Charles Duncan in the vessel *Princess Royal* and James Colnett in the *Prince of Wales* (Halpin and Seguin 1990, 281). After the Hudson's Bay Company moved Fort (later Port) Simpson to its present location, north of Prince Rupert in 1834, nine groups moved to the area of the fort. William Duncan reported there were approximately 2,300 Tsimshian living in 140 houses around the fort in 1857 (Halpin and Seguin 1990, 267–69). It is thought that the Tsimshian were not significantly impacted by the fur traders, but that missionaries had a significant impact on their culture. An Anglican preacher, William Duncan, led a group of converts to a new village on Venn Passage in 1862 and later to New Metlakatla on Annette Island in Alaska in 1887. Metlakatla is the only native reservation in Alaska (Dunn and Booth 1990). By 1873, the Tsimshian that remained at Fort Simpson were Christianised and resembled 'White men' (Halpin and Seguin 1990, 281).

4.5.2. Haida

Haida refers to a group distinguished by their unique language. It is a language isolate, which means Haida is not part of a known language family that has considerable dialectal diversity. The three major surviving dialect clusters are Kaigani in southeast Alaska, Masset in northern Haida Gwaii, and Skidegate in southern Haida Gwaii (figure 4-4) (Blackman 1990; Langdon 1977; Thompson and Kinkade 1990).

At the time of European contact, the Haida lived in 'towns' or villages composed of one or more matrilineal groups. By this time, the Haida had expanded into southeast Alaska from the south, but it is unclear when this happened. From 1850 to 1860, Swanton and Newcombe recorded 126 habitation sites that were occupied by Haida people. Twenty of these were winter villages, the others were resource procurement camps. Haida selected winter village sites based on several criteria including natural protection from storms and enemies, proximity to halibut banks and shellfish resources, availability of drinking water, and adequate beachfront for landing canoes. Houses were built facing the beach nested against the treeline. Ninstints, Kloo, Cumshewa, Chaatl, and Kasaan are winter villages that contained two rows of houses. Above the storm tidemark was a cluster of totem poles including those containing the dead (Blackman 1990, 241).

The Kaigani had a larger inventory of economically useful plants and animals in southeast Alaska than on Haida Gwaii. Southeast Alaska is slightly more protected from storms than Haida Gwaii but has colder winters and warmer summers. However, Haida Gwaii has more abundant halibut and chum salmon and better-quality red cedar (Blackman 1990, 241). Halibut was quantitatively the most important fish. They were caught offshore using a V-shaped hook with line and octopus was a preferred bait. The Masset people in the 1970s identified chum salmon as the most important winter food. The Haida caught sockeye (red), coho (silver), pink (humpback), and chum (dog) salmon in intertidal water in traps set in weirs in streams or with harpoons. The Haida also hunted seals, porpoises, sea lions, fur seal, sea otters, deer, beaver (in Alaska), and caribou (on Haida Gwaii). They also utilised stranded whales and numerous plants and marine invertebrates were collected.

The seasonal cycle varied locally based on available species. Summer was a time of intensive food gathering and processing. Generally, April included hunting birds and collecting their eggs and by late May to early June, people started to harvest sockeye from streams. September was favoured for halibut. October also included hunting ducks and geese and trapping bears and martens. Before the end of November, the chums were smoked and transported back to the winter villages where people remained until the spring. Shellfish was eaten daily and was in the public domain, available to all. Higher valued resources, such as salmon and many berries were owned and controlled by lineages (Blackman 1990, 244–45).

Trade was an important aspect of the Haida economy. The Haida traded canoes, slaves, and shell for native copper, Chilkat blankets, and moose and caribou hides with the Tlingit. The Haida traded canoes, seaweed, and dried halibut for eulachon grease, dried eulachons, and soapberries with the Tsimshian. Trade was conducted between chiefs of equivalent moieties or clans by establishing a bond of brotherhood. The Haida were also known to raid some of the southern NWC tribes for slaves (Blackman 1990, 246).

There were two basic types of Haida houses, and both were constructed of red cedar timbers and planks and were similar in finished form. Type A has seven roof beams. Six beams projected several feet beyond the front. One beam (the ridge pole) spanned the centre of the house. It was installed in two segments. This created an opening in the middle for a smoke hole. This type of house was the most common in central Haida Gwaii but was not found in Kaigani villages. Type B had only four roof beams that did not project outside the house. However, by the late nineteenth century, type A was the most popular with few type B houses found. A carved totem pole normally stood against the house façade, often with the bottom of the pole providing the entrance to the house. Houses were named but may have more than one name based on the crests of the house owner, his economic resources or largesse, events that occurred during house construction, or physical features of the house (Blackman 1990; Langdon 1977).

Haida society was divided into two moieties, Raven and Eagle. Each clan has several lineages or families. Haida lineages differ from Tlingit lineages in that there are no relations between the clans. They are completely fissioned. The Haida lineages trace their origins to several supernatural women of the Raven and Eagle moieties. The Eagle moiety might have been of foreign origin, possibly Tsimshian people. The lineages were named often with reference to the group's origin, special property, or quality. Some lineages, especially those that migrated to southeast Alaska, were divided into sub-lineages. These can be conceptualised as houses and likely reflect a Tlingit influence. The lineages controlled both real and incorporeal property including salmon spawning streams, lakes, trapping sites, patches of edible plants (cinquefoil, fireweed, high-bush and bog cranberry, and crab-apple), stands of cedar trees, bird rookeries, and stretches of coastline. It has been suggested that lineages also owned halibut banks, but in the 1970s the Masset Haida clarified that neither the halibut banks nor the sea were aboriginally lineage properties. Incorporeal property included a repository of names (personal, canoe, fish-trap, house, and spoon names), dances, songs, stories, and crest figures. The lineages were controlled by a hereditary chief who was the trustee of lineage properties. For multi-lineage towns, the highest authority was the 'town master' or 'town mother' which was a title held by the highest-ranking lineage chief (Blackman 1990, 248–51; Swanton 1905, 100).

The Haida practised a class system. Titles were handed down at potlatches to children of high-ranking lineages by their parents. People who did not have potlatches held in their benefit were not high-ranked and were less successful as high-ranked individuals in economic endeavours. Ceremonies were directly related to rank in Haida society. Slaves were war captives and their offspring (Blackman 1990, 252).

The first recorded contact with Europeans was in 1774 by Juan Perez at Langara Island on Haida Gwaii and on Dall Island in Alaska. European and American mariners traded extensively with the Haida throughout the late eighteenth and early nineteenth centuries. The Haida traded sea otter pelts for manufactured goods. Potato cultivation was introduced by traders. By 1825, the Haida were growing large quantities of potatoes and trading them with the coast Tsimshian and later the Hudson's Bay Company. The Haida were regarded as skilled traders and were able to increase the price they received for their pelts. It is unclear the full effect that the European traders had on the Haida, but introduced diseases significantly decreased the population. The last traditional house and frontal totem-pole raising occurred in the winter of 1881 (Blackman 1990, 255–57; Gibson 1992).

4.5.3. Tlingit

The Tlingit share a common language and customs (de Laguna 1990: 203). Tlingit is a Na-Dené language. It is similar to Eyak and Athapaskan (figure 4-3). Tlingit is a homogeneous language with only mild dialect diversity. The dialects of Tlingit are Gulf Coast, inland, northern, and southern (Thompson and Kinkade 1990, 31). Recent research indicates that Ket, a Yeniseian language from central Siberia, is related to Tlingit and Eyak (Kari and Potter 2010).

There are three major groups of tribes (figure 4-4): The gulf coast, the northern Tlingit, and the southern Tlingit. There also are inland Tlingit, including groups in the Yukon, but this review focuses specifically on the Coastal Tlingit. The Gulf Coast Tlingit include the Dry Bay and after contact the Yakutat. The Yakutat are a group of Eyak that joined the Dry Bay Tlingit. The northern Tlingit include the Hoonah, the Chilkat-Chilkoot, Auk, Taku, Sumdum, Sitka, and Hutsnuwu. The southern Tlingit include the Kake, Kuiu, Henya, Klawak, Stikine (or Wrangell), Tongass, and Sanya (or Cape Fox). The tribes (and in figure 4-4) are local communities consisting of several clans united by propinquity and intermarriage. Tlingit society is organised in matrilineal clans of two exogamous moieties. The clans held the territorial rights, and the tribes lacked chiefs or councils (de Laguna 1990; Langdon 1977).

Tlingit tribes had at least one principal village that was occupied in the winter. In the summer, it was deserted when the families left for hunting and fishing camps. Additional settlements were formed when immigrant groups arrived or if an old village was being abandoned or divided due to disputes, war, or disease. The preference for village locations is within sheltered bays with a sandy beach from which there is a view of approaches with nearby salmon streams. Proximity of hunting areas, berry patches, clam beds, freshwater, good timber, or special resources (such as halibut deeps, sealing grounds, or trails to the interior) also was important factors in site selection. Within the village, houses were arranged in a row facing the water. In front of the houses were canoes under mats, fish-racks, garbage disposal areas, and totem poles. The graveyard was usually at the end of the row of houses or on an island opposite the settlement. Behind the houses were smokehouses, caches, steam bath huts, women's menstruation huts, and childbirth huts (de Laguna 1990, 206–7).

Seasonal rounds varied not only geographically, but different families might choose to follow different pursuits thus providing a more reliable and diverse resource base. Generally, the spring included hunting and trapping bear, marten, mink, beaver, sea otter, fur seal; catching halibut in deep water; collecting shellfish, seaweed, and herring spawn; and, gathering eulachon and processing it for oil. The first chinook (king) salmon runs began in April and May and much of the summer was spent catching and curing salmon. In addition to collecting berries, some Tlingit were occupied with harbour seal hunting, warfare, and slave raids to the south. The fall was a time for sea otter hunting, but catching salmon, smoking it, and picking the last of the berries were often more important. From collecting at the salmon streams, many people went straight to hunting because the animals were fattest at that time of year. By middle October, many families were established in their winter villages. The winter months were the time for ceremonies, trading trips into the interior, and craftwork (de Laguna 1990, 206; Oberg 1973).

Tlingit houses were rectangular with low-pitched gable roofs. They accommodated around 40 to 50 people, including approximately six families. The centre was excavated and planked to form a central fire pit. There were one or more wide wooden platforms, the highest of which were used as family sleeping areas. These sleeping areas were partitioned with wood screens, mats, or boxes. Under the platforms in dark holes was where some girls spent their puberty confinement or accused people were imprisoned until they confessed. Steam baths were also under the floor accessed by a trapdoor near the fire. The front entrance was an oval set above the usual winter snow level. The only other opening was a smoke hole with a screen that could be tilted against the wind. Wall planks were either vertical or horizontal. The four main house posts were carved and painted with totemic and ancestral figures. Houses, their posts, and totem poles were named and were owned by the lineage or clan houses. The houses built in a hurry were made of spruce or cedars bark (de Laguna 1990, 207; Oberg 1973).

Because of the limited distribution of red cedar, most Tlingit primarily made their canoes from large spruce and paddles from yellow cedar. The southern Tlingit had access to red cedar and Tlingit's preferred Haida canoes for their size. They were involved in intertribal trade with Eyak, Tsimshian, Haida, and subarctic Athapaskan tribes. Trade was usually between 'partners' who were members of the same moiety but from different clans or between 'brothers-in-law' who were members of opposite moieties, or between 'fathers- and sons-in-law'. Clan leaders owned the trails into the interior. They forbade direct contact between 'whites' and interior Athapaskan (de Laguna 1990, 208–9).

Tlingit oral history indicates that clans migrated northward from the Tsimshian peninsula and that later other clans moved down the Skeena, Nass, Stikine, and Taku Rivers at a time when they were blocked by glacial ice (de Laguna 1990; Langdon 1977). 'Human occupation of the Tlingit coast could have begun 10,000 years ago' (de Laguna 1990, 206). In the eighteenth century, the Tlingit were expanding into the Gulf Coast of Alaska, possibly due to the displacement of southern Tlingit groups from the Haida's northward migration into southeast Alaska. The northern Tlingit territories likely were occupied between AD 1200 and 1700 when the glaciers were smaller than they are now, but archaeological remains have not been discovered.

Tlingit social, political, and economic organisations were based around the two moieties: Raven and Wolf (sometime Eagle in the north). Each moiety included approximately 30 clans that were further subdivided into lineages or house groups. Memberships in all three groups were matrilineal. Moieties never met and were not social groups. They simply arranged people into opposites. These opposites were required to perform important social and ceremonial services (Boas 1982, 376). The Henya had an Eagle clan that was outside the other two moieties. Tlingit clans and houses possessed territories and the rights to all game, fish, berries, timber, drinking-water, and trade routes; house sites in the winter village; and, totemic crests, decorations of houses, heirloom objects, and personal names. The chiefs were trustees and administrators of these properties (de Laguna 1990, 212–13; Oberg 1973).

European contact is first recorded in 1741 by the Russian explorer Alexei Chirkov. The Russians lost two boats in the encounter. Later, there was Bruno de Hezeta in 1775 from Spain who explored Klawak, Sitka, and Hoonah territory, and because of this contact, the Sitka Tlingit were infected with smallpox. Many Russian, British, and American vessels visited the Tlingit during this period. In 1799, Aleksandr Baranov established a fort at Sitka, which was captured by the Tlingit in 1802. In 1804, the Russians reclaimed the area and established Novo-Arkangel'sk (New Archangel), later Sitka. In 1808, this became the headquarters of the Russian-American Company. Alaska was purchased from the Russians in 1867, despite Tlingit demands that the Russians did not own Alaska. In 1971, the Tlingit and Alaska Haida formed the Sealaska Regional Corporation and village corporations in 10 villages, under the provisions of the Alaska Native Claims Settlement Act. This reinvigorated a revival in Tlingit culture. Since 1970, Tlingit art (singing, dancing, and woodcarving) has flourished along with potlatches and other traditional practices (de Laguna 1990, 223–26; Gibson 1992; Worl 1990).

4.6. Site Location Descriptions

In addition to the general descriptions of where Native people lived in southeast Alaska prior to the time of Euro-American contact, a few quotes illustrate the factors underlying the selection of winter village locations. These are the types of sites that would be the most visible archaeologically, especially house structures with excavated pits similar to the ethnographic plank houses.

Meares (1790: 109), an English ship captain, noted that Nootka was located in a protected harbour or bay. This does not indicate the location of their freshwater source. On 28 June 1788, Meares (1790, 146) notes that the village of Wiscanaish moved from close to the sea to the inner port. The next day they came across another group that is identified as being under Wicanaish's domain. Their village located along an open stretch of coast (Meares 1790, 152). When he entered the Strait of Juan de Fuca, Mears' saw villages on high banks close to the sea (Meares 1790, 157). 'On the mainland there are large and populous villages, well-watered by rivulets, where great numbers of salmon are taken, which, when prepared properly, constitutes a principal part of their winter's food' (Meares 1790, 172).

Mary Gormly (1971, 163) says that the Tlingit of Bucareli Bay had their houses in defendable positions, usually on steep and rugged hills that were approached by a ladder. Moss and Erlandson (1992) discuss this archaeologically. Forts are seen in the Late Period and ethnographic period as warfare became more predominant.

De Laguna worked with the Angoon Tlingit investigating both archaeology and ethnography. She states:

> 'Most of the village sites were small flats, cramped between the beach and the steep hillside, and where space permitted the houses were ranged in a line just above the water. Sites for settlements were chosen more for a good landing beach for canoes then for convenient access by trail to inland hunting or trapping grounds.

Summer villages and camps might be far up the bays near the salmon streams, but for the winter villages of permanent houses the people prized a view of the more open water across which the canoes of their friends or their enemies might be seen approaching'. (de Laguna 1960, 30)

Archaeologically, Moss (1992) describes sites being located in proximity to food sources and on banks or mouths of salmon streams. These would be the summer camps not the winter villages. These site location descriptions, in addition to the ones from the main text, provide a means of creating the archaeological site high potential model.

More generally, Easton (1993, 3) states:

> 'If we accept the assumption that most coastally adapted populations live close to the existing shoreline of the day, as they do today and have in the documented past, then rising RSL [relative sea levels] would inundate evidence of ancient shoreline occupations.'

These descriptions were incorporated qualitatively into the model weighting process. This includes increasing the model weight to coastlines that demonstrate medium sinuosity to locate the bays where people would have preferred to live. Higher weights were assigned to streams and lakes for inland fishing grounds and possible routes into the interior. These locations are incorporated into the decisions regarding the weight of variables within the models. It is a basic assumption of this research that ancient habitation sites in southeast Alaska would have been near the coast.

5

The GIS Model

5.1. Introduction

A GIS model was designed with the intention of locating high potential areas for the occurrence of non-site and sites where past people would have exploited resources and the environment. The majority of the inputs into this model are environmental (slope, aspect, and distance from freshwater). Other variables are more subjective, such as distance from resource patches or between known archaeological sites (because the locations are incomplete). The time depth of the model must be incorporated into its development. Many models and maps are created utilising the modern land surface, despite knowledge of temporal variability of the environment. Because the focus of this research is from 11,000 to 16,000 cal BP, resource patches are difficult to identify and locate. Consequently, resource patches are not directly applicable variables for this model. However, land-sea level relationships can be reconstructed to appropriate temporal intervals, or 'time slices', using modern bathymetry and sea level history. Since the majority of the resources exploited by people along the northern NWC were maritime resources, the coastline is an appropriate approximation for resource locations. The data sources used to create the palaeo-landscapes and the models are discussed below, followed by the method for creating the models.

5.2. Data

The vector (point, line, and polygon) data were downloaded from the Alaska State Geo-Spatial Data Clearinghouse (ASGDC) (State of Alaska, n.d.). This includes hydrology (including rivers and lakes), communities, modern glacial extent, and coastlines. These data sets were used for visualisation and data exploration. The geologic maps for the region were also downloaded as vector data from ASGDC. This included polygon data of Gehrels and Berg's 1992 map of the exposed geology of southeast Alaska (figure 3-9). These data were used to characterise the geology of the land within the survey areas (see chapter three: Study Area – Shakan Bay). The Canadian data was downloaded from GeoBase. GeoBase mainly provided hydrology shapefiles and topographic maps.

Nautical charts were downloaded from the NOAA website. These were used to test the accuracy of the DEM (digital elevation model) produced for this project, to analyse sea floor composition, and for data exploration. From the USGS publications website, the shapefiles for Carrara and colleagues (2003) publication were downloaded. Raw GIS files were also provided by the Tongass National Forest courtesy of James Baichtal. These data were used mainly for exploration and mapping purposes, not for data analysis.

5.2.1. Bathymetry and DEM

Bathymetry is the topography of the seafloor. Unlike land features, there are few sources to downloaded compiled elevation models of the seafloor. Initially, a one-arc second grid file was obtained from the NOAA website. This file later proved to have too many data errors to be useful for analysis. Data was downloaded from the hydrology section of the NOAA website. This included sounding data from early surveys, multibeam sonar surveys, and other surveys within the region (Monteleone 2013). The geographic extent of the study region was used to limit the search.

All of the data were downloaded as xyz point files (ASCII). These files were processed using a python script that converted the data to point shapefiles in UTM Zone 8 North. The steps included converting the files to comma-delimited files, adding a header ('latitude, longitude, z'), generating a xy extent, then converting the xy extent into a shapefile, and finally converting the NAD83 shapefile into UTM Zone 8 North, when necessary. After all of the files were in the same format, they were merged in ArcGIS into a single shapefile.

Other bathymetry data were purchased with NSF (OPP #0703980) funding from Scientific Fishers Inc. in Anchorage, Alaska. This included bathymetry points and coastal points in the shapefile. The coastline of this data was actually at nine meters elevation, associated with high tide. The NOAA data were merged in ArcGIS with the coastal and bathymetry data from SciFish Inc. to form the final bathymetry data used to generate the DEM.

The land topography was downloaded from the USGS website using Earthexplorer. Four datasets were used: [1] National Elevation Dataset (NED) (Gesch et al. 2002) at one-meter resolution; [2] GeoSAR Interferometric Synthetic Aperture Radar (IFSAR) with a resolution of five-meter; [3] IFSAR with a resolution of 2.5-meter; and, [4] the multibeam data collected in 2012 in Shakan Bay.

To create the final seamless DEM including both the bathymetry and the land topography, the point files from the bathymetry and the land were merged to form a large (approximately 40 million-point) data file (figure 5-1). This file was then converted to a raster using Inverse Weighted Distance (IWD) in ArcGIS's 3D Analyst at one-meter resolution. A minimum of four points was used per cell with a variable search radius and a power of one. The final DEM has elevations ranging from -422.5 to 1,019 meters. This version of the DEM was created for the smaller study area. Originally, a five-meter resolution continuous DEM was created for the study region using the same methods.

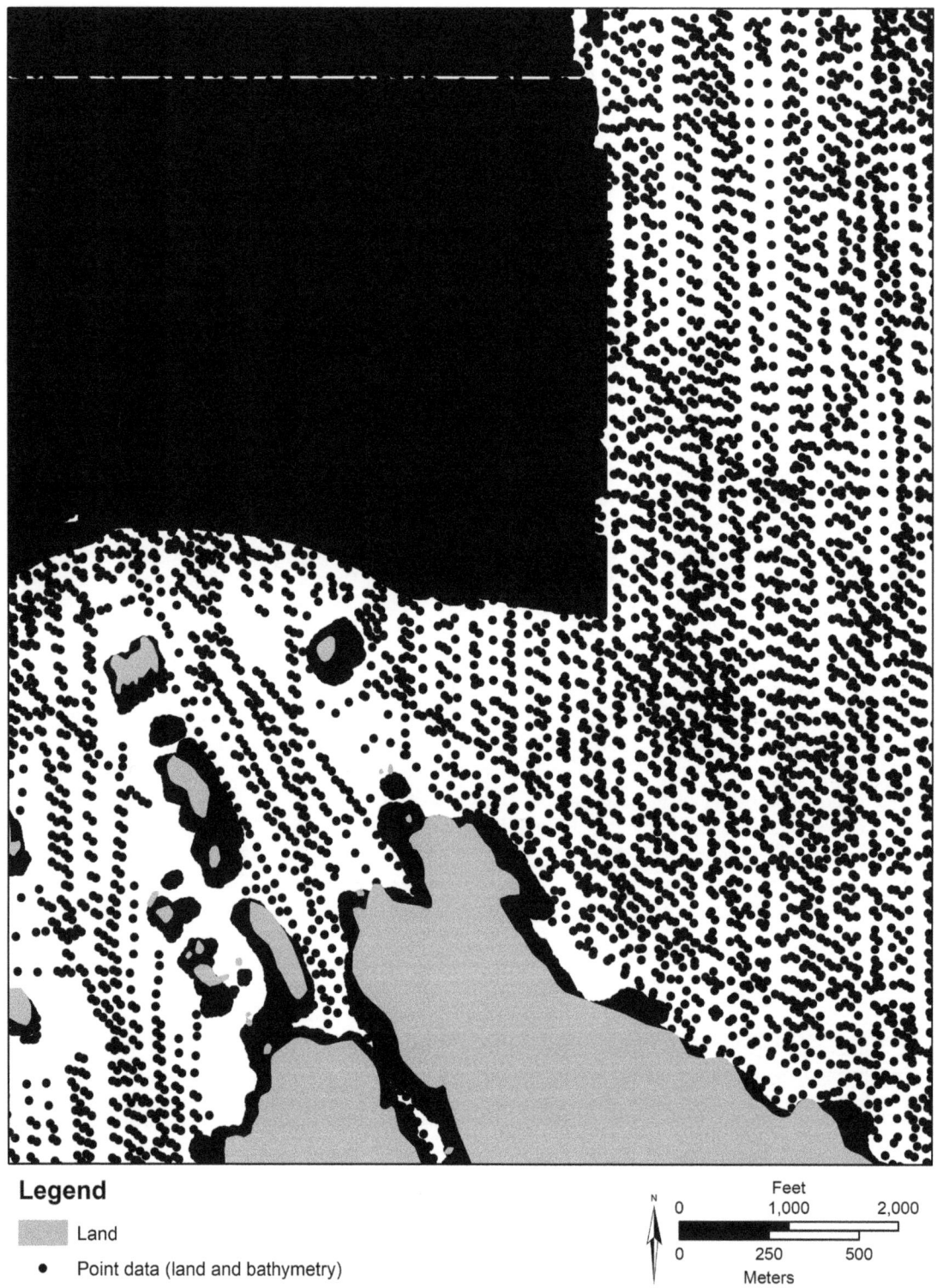

Figure 5-1. Distribution of data points included in the bathymetry DEM of the study area.

The DEM was tested using multiple methods. First, the data were compared to NOAA charts for the region (figure 5-2). The contour lines used by NOAA (5, 10, 25, and 50 fathoms) were converted to meters (9.144, 18.29, 45.72, and 91.44 m) to aid visual comparison between the NOAA charts and the DEM. Generally, the contour lines match between the two data sets. The differences between the contour lines are in places where there are no NOAA data. Although the DEM and NOAA charts are generally consistent, this could suggest that the larger dataset used to compile the DEM may be of higher resolution.

Another test of the DEM was to compare the method used, IDW, with another method to create the seamless surface, spline. A Spline DEM (ArcGIS 10 - 3D analyst tool) was generated at one-meter resolution. Figure 5-3 A is the Spline DEM using the same colour ramp as figure 5-3 B. Figure 5-3 C was generated by subtracting the Spline

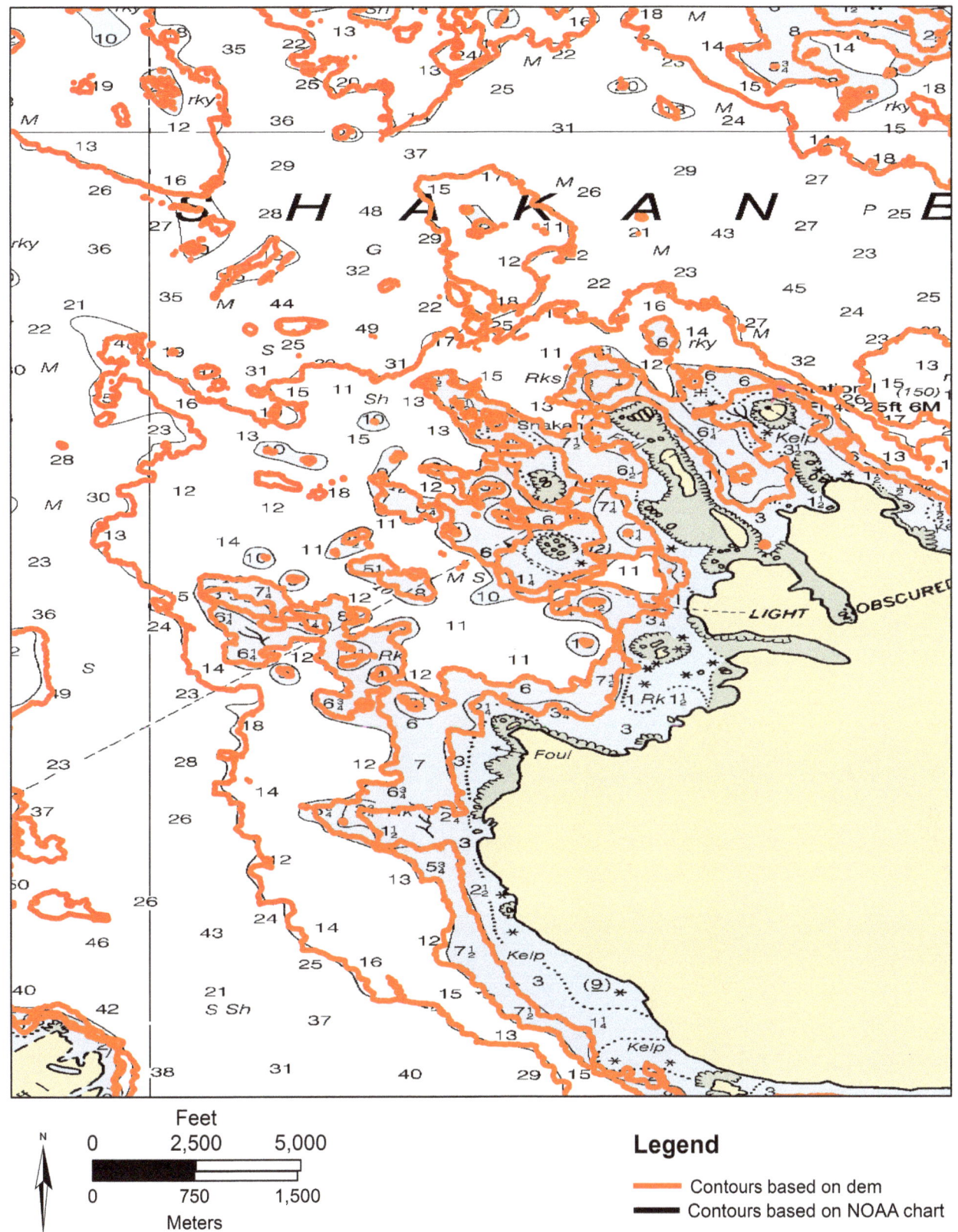

Figure 5-2. DEM generated for this project over NOAA chart with matching contour lines (black lines NOAA, red lines DEM).

DEM from the IDW DEM and taking the absolute value of the results. A coarse (25-meter resolution) version of DEM with no bathymetry data was included for visual reference. The mean value, before absolute value was run, is -0.115 meters, indicating that the two DEMS are similar.

Therefore, the IDW DEM was determined to be adequate for this study.

Figure 5-4 represents the difference between the multibeam data collected in May 2012 and the IDW

Uncovering Submerged Landscapes

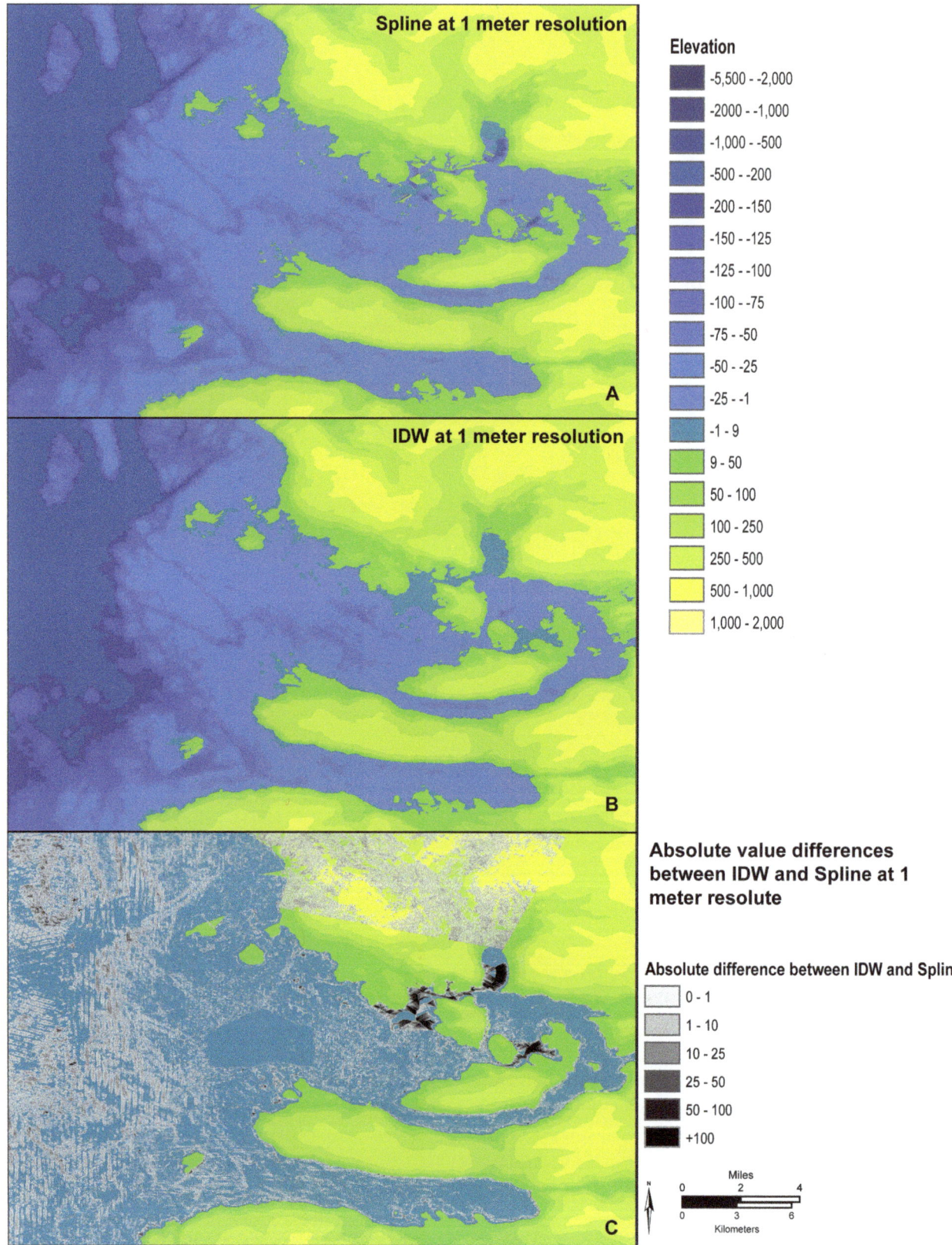

Figure 5-3. Comparison of Spline and IDW DEM methods. A) Spline DEM. B) IDW DEM. C) The absolute value of IDW minus Spline DEMs. The comparison was conducted between the IDW and a Spline version of the DEM to determine if there would be a significant difference in results between the IDW and a Spline version of the DEM.

DEM. The multibeam data was subtracted from the DEM using raster math. The multibeam and the DEM have some variability (mean 2.19, standard deviation 1.8) between their interpretations of the seafloor. The accuracy of the one-meter resolution DEM is significantly higher than that of the five-meter resolution where the mean was 5.24 meters and the standard deviation was 27.24 meters. The maximum variability is 32 m. The multibeam values

The GIS Model

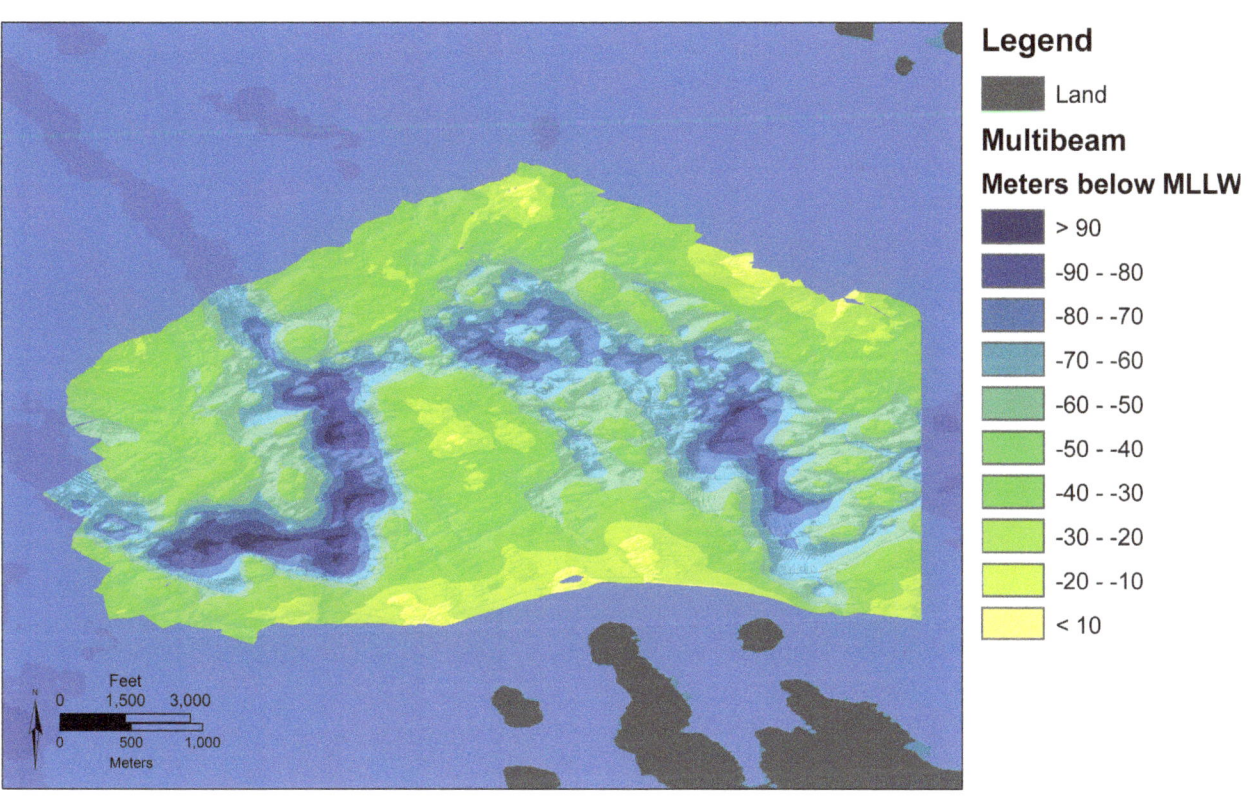

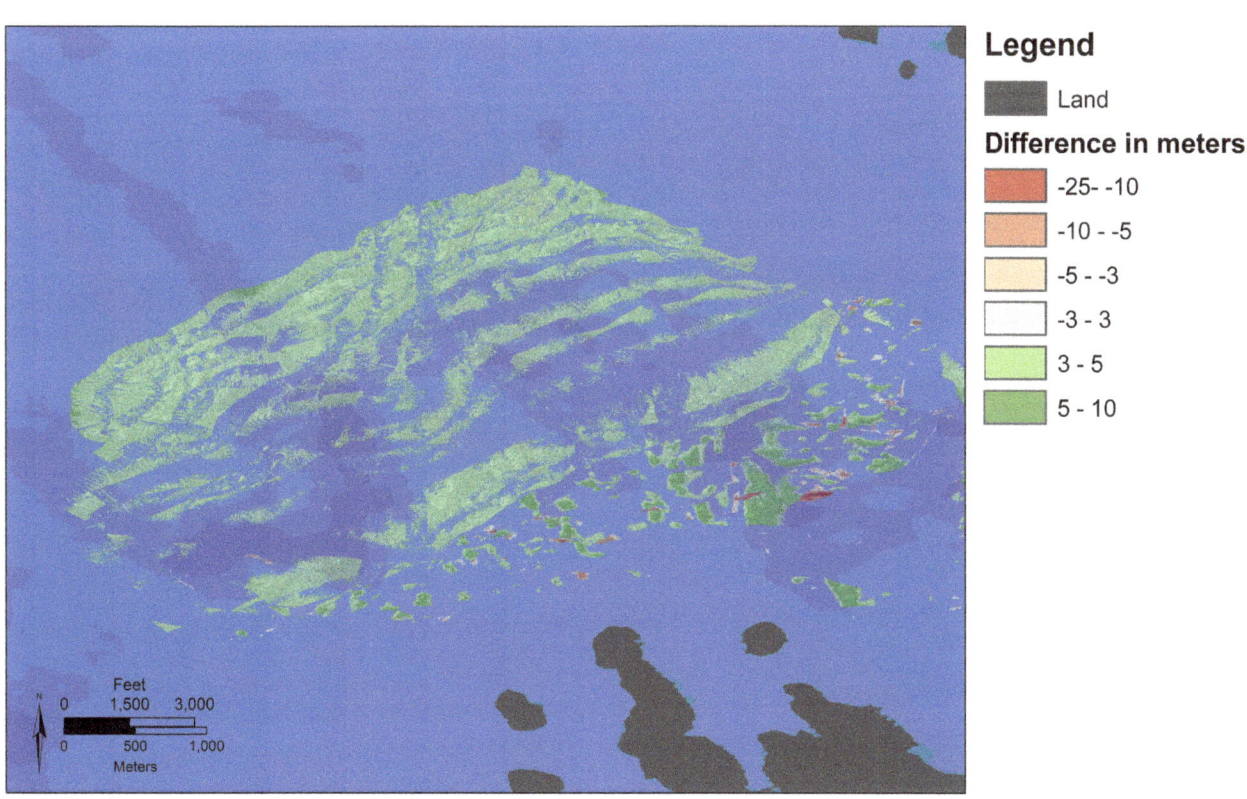

Figure 5-4. Multibeam. A) Multibeam data collected for Shakan Bay in 2012. B) Difference between DEM and multibeam at 1.5 m resolution.

appear to be below the DEM in the south-southeast and above the DEM through the majority of the area. This discrepancy corresponds, partially, to when the surveys were conducted. Hence, a possible explanation is that there is an error in the tidal corrections or the velocity of sound corrections. This can be corrected using different post-processing methods and more careful data collection.

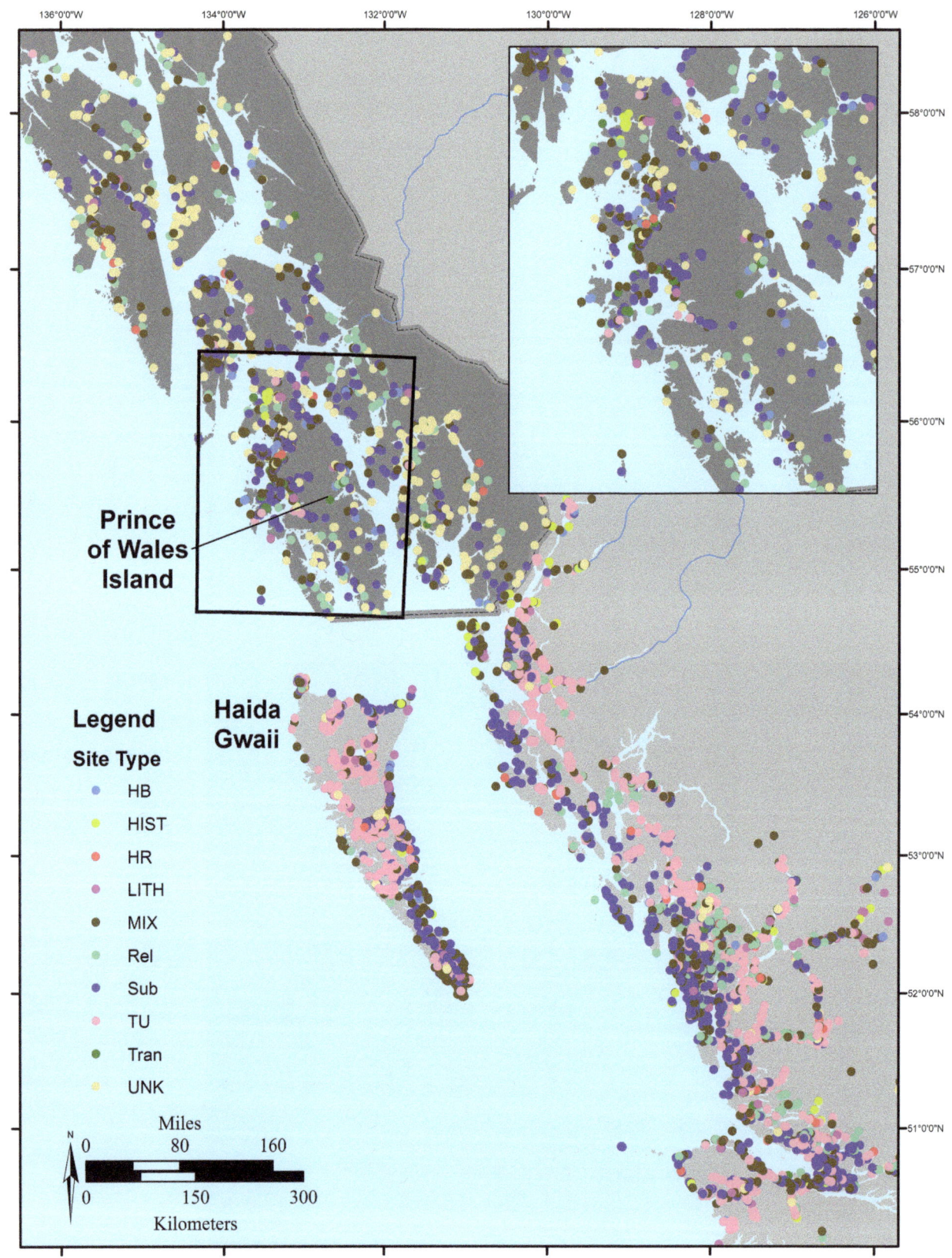

Figure 5-5. Map of archaeological sites in the project database (see table 5-1 for site types).

5.2.2. Archaeological Site Data

In total, there are 7,470 archaeological sites in the project's database (figure 5-5). These data were provided by the Alaska SHPO and BC Archaeology Branch in January 2009. These sites range from the northern half of Vancouver Island (50° 3.3' N) to just north of Juneau, Alaska (58° 53.8' N). The BC data followed forest district boundaries and includes Haida Gwaii, North Island-Central Coast, and North Coast Forest. The Alaskan data encompass everything south of Juneau in Southeast Alaska. New site information was downloaded from the Alaska Heritage Resource Survey website in fall 2018. Some sites were not included in the original dataset as they were historic others were more recently discovered.

All sites were manually assessed for analysis. Table 5-1 has the ten identified site types and their frequency in the total database of 7470 sites, the study region of 931 sites, and the study area with 42 sites. Habitation sites (HB) include sites identified as forts, battle site, midden,

Table 5-1. Frequency of archaeological site types within the entire database, study region, and study area. New sites are all in the study area and were not part of the model production

Site Type		Total (in database)	Study region	Study area	New (not in model)
Habitation	HB	67	31	0	0
Historic	HIST	84	10	9	9
Human Remains	HR	99	7	0	0
Lithic	LITH	158	12	2	2
Mixed	MIX	1325	225	6	1
Palentological	PAL	4	4	4	4
Religious	Rel	432	74	4	1
Subsistence	Sub	2267	345	8	2
Transportation	Tran	40	29	1	1
Traditional Use	TU	2472	10	1	1
Unknown	UNK	522	184	7	1
TOTAL		7470	931	42	22

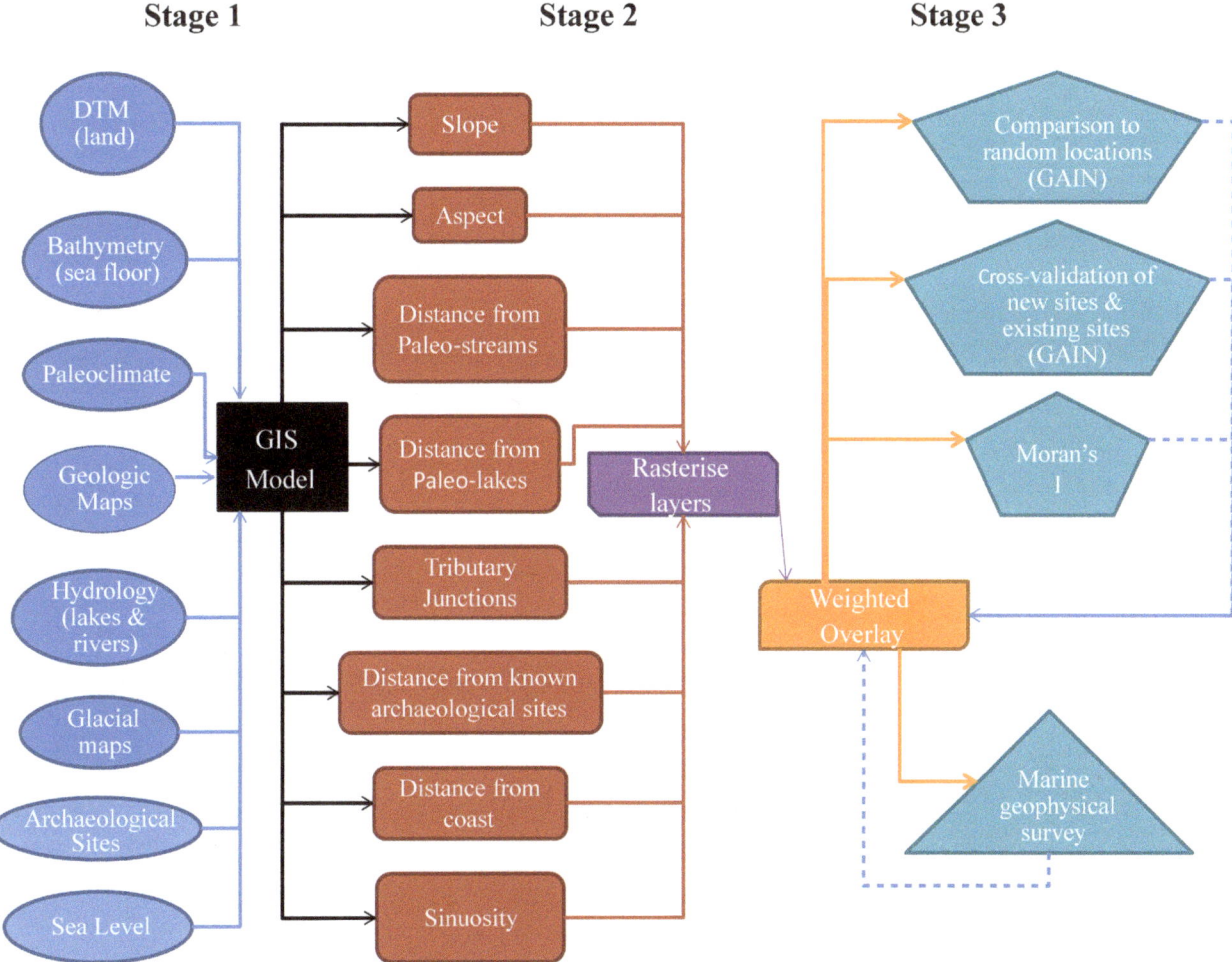

Figure 5-6. Flowchart of three stages for developing the model.

activity area, settlement, camp, rock-shelter, depressions, village, or house depression. Historic sites (HIST) include sites identified as historic structures, cabins, World War II structures, mines, and farms. Human remains sites (HR) include sites identified as burials. Lithic sites (LITH) include sites identified as quarries, lithic scatters, lithic remains, and rock alignments/formations. Mixed sites (MIX) list more than one of the identified site types. Palaeontological sites (PAL) do not include archaeological material. They were only added to the study area. Religious sites (Rel) include sites identified as petroglyphs, pictographs, mythological or oral tradition, and carved ceremonial trees. Subsistence sites (Sub) include sites identified as fishing weirs or traps, middens including shell or bone, and faunal remains. Transportation sites (Tran) include sites identified as canoe landings, canoe remnants, canoe runs, and a shipwreck. Traditional use sites (TU) include sites identified as culturally modified trees (CMTs) including bark stripping and pitch trees. Either unknown sites (UNK) do not have a detailed description or the description is so brief it is unclear what was found, such as 'isolated find'. Assuming a statistical sample population minimum of 20, there are enough sites in each category for statistically valid analysis for each site type except for lithic, human remains, traditional use, and historic (because historical sites were not provided for the study region). Site types that would not be useful for the modelling process are traditional use, historic, and human remains. Religious sites may be useful if petroglyphs or pictographs could be identified with an ROV, but the locations are likely not useful for modelling.

The descriptive information for many of the archaeological sites is limited, but locational information is nearly always present. Only 254 of the sites in the Alaskan database include radiocarbon dates. Some additional radiocarbon data were obtained from published sources and the Canadian Archaeological Radiocarbon Database (canadianarchaeology.ca). All archaeological site types found in southeast Alaska were incorporated in the modelling process and statistical analysis. In total, 909 known archaeological sites within the study region were analysed.

5.3. The Three Stages of the Model

The archaeological high potential model was developed in three stages (figure 5-6). Stage one is the GIS and includes gathering data, formatting it, and incorporating it into an ArcGIS database. Stage two can be divided into two types of products. Intermediate products are the raster datasets derived from the Stage one data. These include

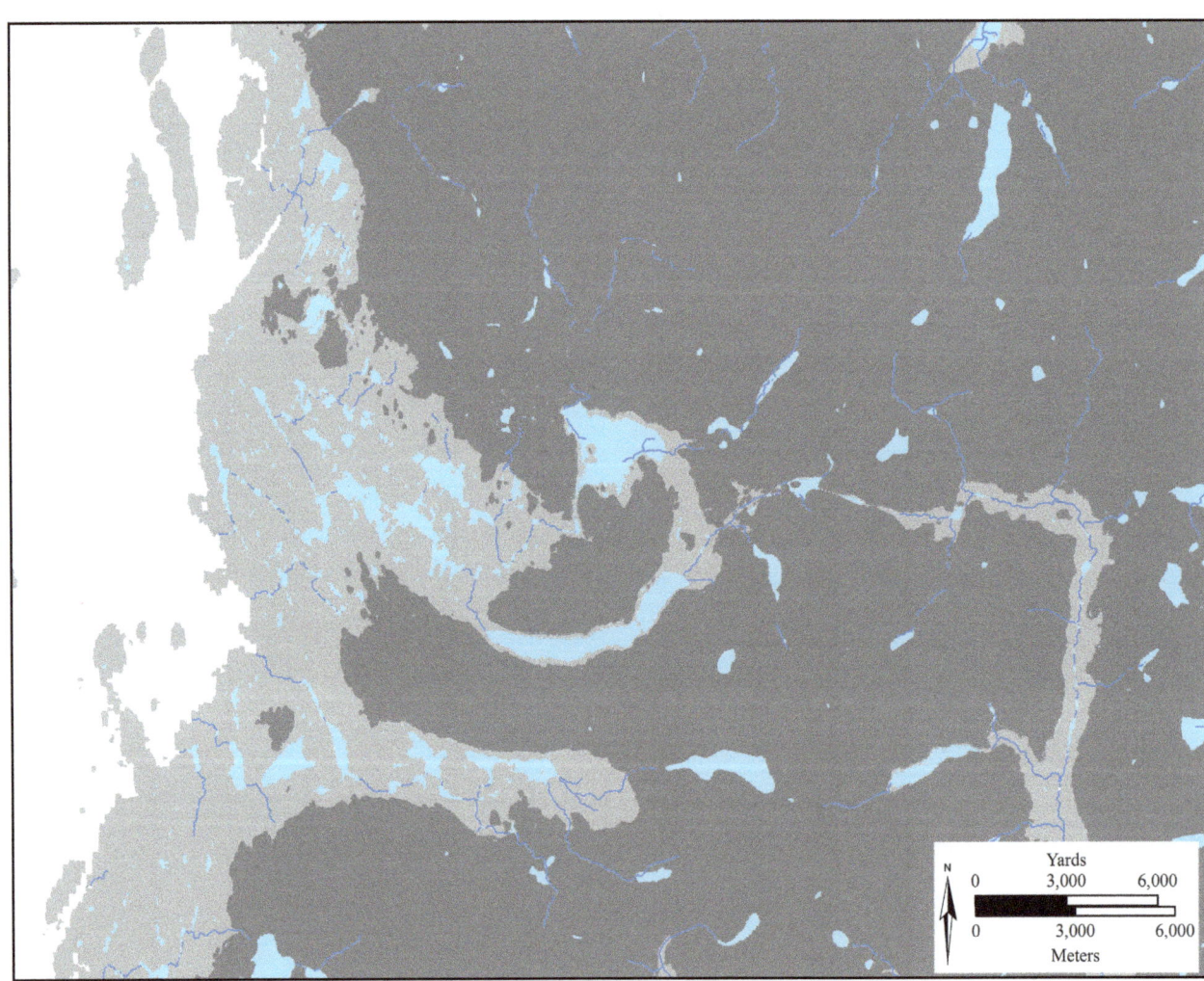

Figure 5-7. Lake and streams reconstructed in Shakan Bay study area for 16,000 cal BP, darker land is modern.

Table 5-2. Results of Welsh two-sample t-test for model variables between archaeological site locations and random locations. Values in bold are where t-test null hypothesis cannot be rejected. Sites added to study region were not included in analysis (see table 5-1)

	General Statistics										Welsh Two Sample T-test				
	Min sites	Max sites	Mean sites	Median sites	Number of sites	Min random	Max random	Mean random	Median random	Number of random locations	95% Confidence Interval of the Difference		t	df	p-value
											Lower	Upper			
Slope	0.0	60.0	7.3	5.0	903	0.0	57.0	14.6	12.0	532	6.10	8.48	12.04	830.84	0.00
Aspect		352.0	155.3	166.0	903		357.0	160.1	158.0	532	-6.49	16.08	0.83	1028.80	0.40
Distance from site	30.8	28063.7	1520.9	698.0	909	79.1	54271.3	9646.6	6202.6	1000	7490.73	8761.67	25.09	1095.40	0.00
Distance from lake	0.0	2349.9	183.0	110.0	909	0.0	54028.2	4828.0	907.0	1000	4062.32	5229.45	15.62	1000.20	0.00
Distance from stream	1.5	2349.9	524.5	380.4	909	1.0	30585.9	1476.9	2695.9	1000	784.15	1120.58	11.11	1070.30	0.00
Distance from tributary	8.5	31730.7	1348.8	462.5	909	12.1	82151.3	7458.8	2268.0	1000	5291.38	6928.68	14.65	1075.20	0.00
Distance from coast	1.5	16603.0	583.8	166.0	909	0.6	53755.2	6224.5	2695.9	1000	5072.63	6219.77	19.32	1046.70	0.00
Nearest Sinuosity value	0.0	387.4	4.7	0.0	909	0.0	768.0	7.5	3.6	1000	0.55	5.04	2.45	1421.30	0.01

slope, aspect, coastal sinuosity, and the distance from coast to freshwater and other archaeological sites. The final products are weighted to develop overlays that depict the high potential areas. The third stage of the model was testing. It was conducted in two ways. The first (stage 3a) was statistical testing (cross-validation using Kvamme's Gain, high/low clustering or Getis-Ord General-G, and spatial autocorrelation with Moran's I). Stage 3a, or the statistical validation, is discussed in detail in Chapter six: Analysis of the model. The second testing method (stage 3b) was field testing using marine geophysical survey. The survey included multibeam sonar, side scan sonar, sub-bottom profiler, an ROV (remotely operated vehicle), and a sediment sampling unit (van veen grab sampler). The results of the marine geophysical survey are presented in chapter seven: field-testing the model. The model incorporates an iterative process where model values are adjusted to reflect testing results and then retested. The version of the model discussed in this volume is the fourth iteration.

For the study area, models were generated for eleven 500 cal BP year intervals spanning the late Pleistocene, between 11,000 to 16,000 cal BP. The previous iteration of the model, 10-meter resolution study region, was analysed between 10,500 and 16,000 cal BP. As 10,500 cal BP was actually above modern sea level, this time-slice was removed for the latest iteration. The land-sea level relationships were spatially depicted using a zero m elevation to represent the modern coastline using the one-meter DEM. Intermediate products, such as slope, aspect, and buffers of the polygon features, were created at two-meter resolution. The final product was also created at two-meter resolution. Other iterations of the model included a 50-meter resolution version of the model for the northwest coast (the area depicted in figure 5-5) and a 10-meter resolution of the model for the study region. The 50-meter NWC model was only created for modern sea level. The focus is on the two-meter resolution model for the study area centred on Shakan Bay. However, the statistical analysis of variables was conducted on the 10-meter resolution for the study region, due to the limited archaeological sample size in the study area.

5.3.1. How This Research Fits with Similar Site Discovery Models

Two models for intertidal/submerged site discovery have been clearly presented in the literature: [1] the NWC specific model recently published by Mackie, Fedje, and McLaren (2018) and [2] the Danish Model. In this NWC model, researchers combine high resolutions sea level histories with an accurate digital elevation model

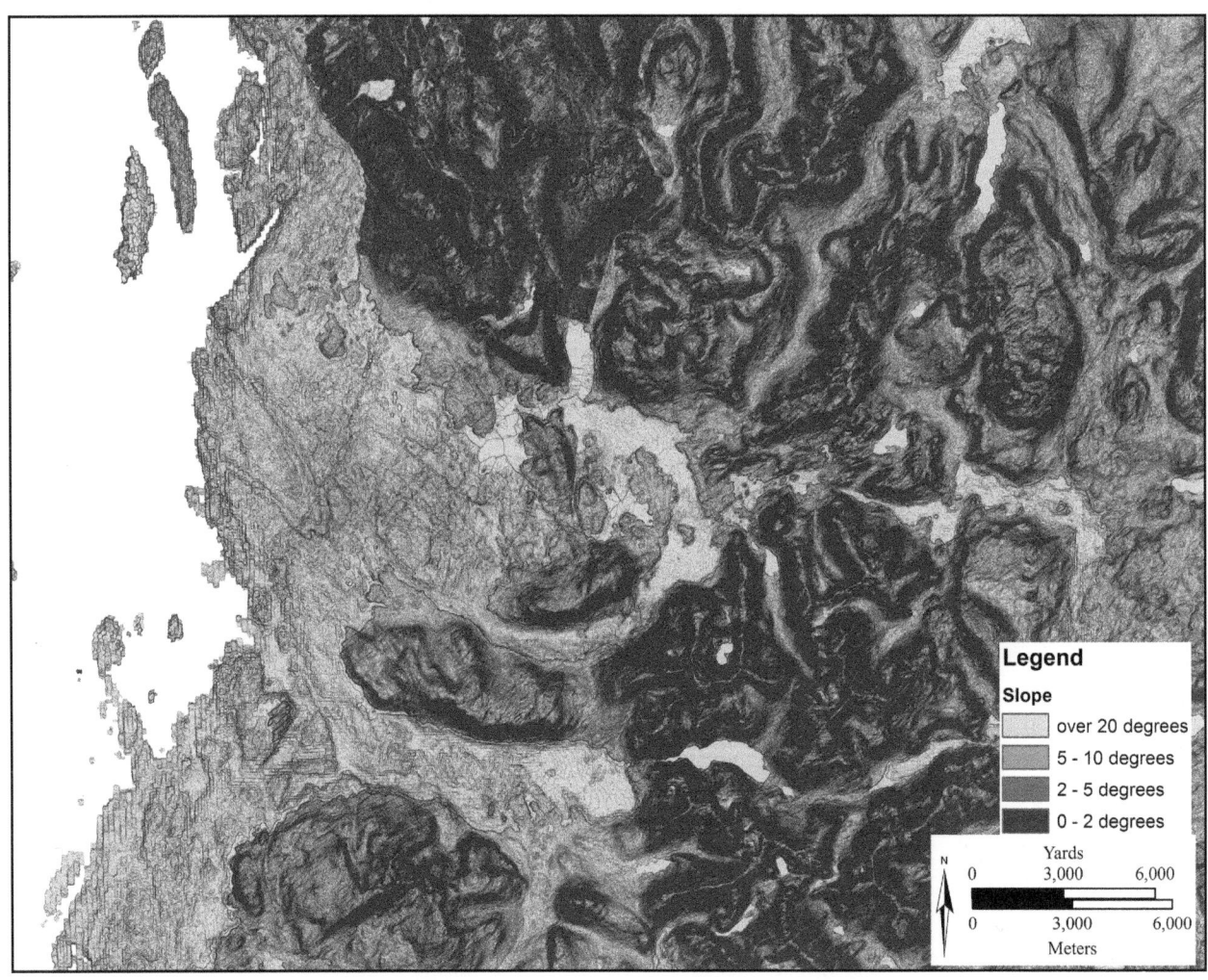

Figure 5-8. Reclassified slope raster for Shakan Bay study area 16,000 cal BP (two-meter resolution).

The GIS Model

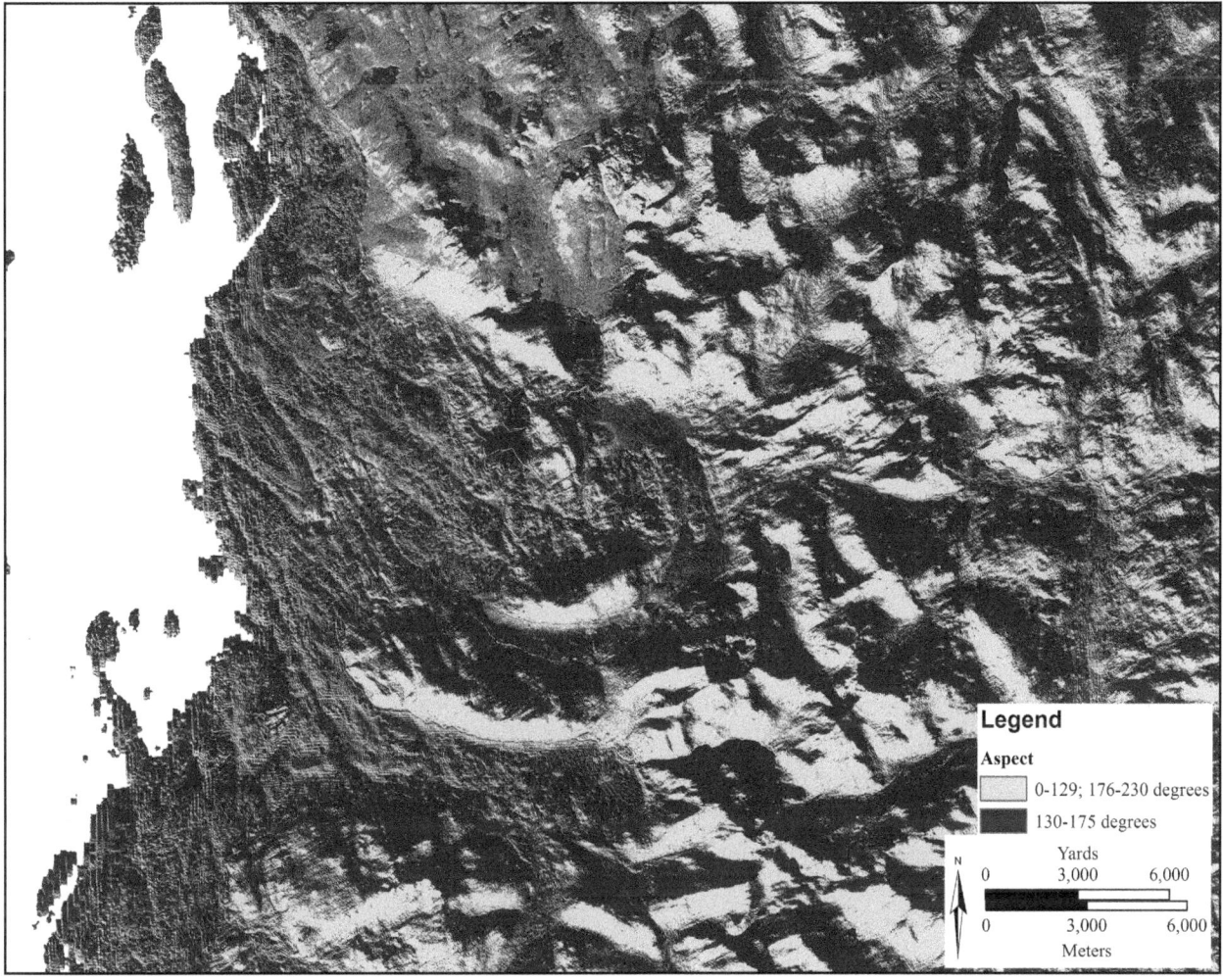

Figure 5-9. Reclassified aspect raster for Shakan Bay study area 16,000 cal BP (two-meter resolution).

to target sites on palaeo-coastlines. Though the sea level for the study area and study region could use more work, Mackie, Fedje, and McLaren's (2018) model is essentially the process followed by this research, in hindsight. The Danish Model (Benjamin 2010) for submerged ancient site discovery (discussed in chapter two), which has six phases of application.

The Danish Model, phase I defines the general background for the region including archaeology, geography, geology, and geomorphology. This phase was conducted through independent study classes during the author's PhD and independent research. Phase II is ethnographic background and historical research based in part on interviews in 2007 (CU Seed Grant) by the PI (E. James Dixon). Chapter four includes a review of the ethnographic literature for the region. Phase III is the evaluation of maps, charts, satellite data, and aerial analysis and plotting locations of known archaeological sites. For this research, the development of the high potential model was determined to be practical and feasible based on these criteria. Phase IV is the observation of potential survey locations which is the geophysical survey conducted to test the model. Phase V is marking of theoretical sites with GPS and diving to investigate. As our survey areas were too deep, an ROV and van veen grab sampler was used. Phase VI is post-fieldwork analysis. This is the statistical evaluation of the model and the interpretation of results. This project specifically focuses on Phases III through VI.

5.3.2. Stage One – Inputs for the Model

There are eight inputs in stage one of the model. The DTM (digital terrain model) or land and bathymetry data were developed from point data. Archaeological site data were compiled for the northern NWC. Palaeoclimate was used as an input in developing the hydrology (lakes and rivers). Geologic maps, glacial maps, and palaeoclimate

Table 5-3. Coastal sinuosity ranks in relation to the location of recorded sites

Sinuosity Value	Rank
< 1	Low
1.01 - 4	Medium
4.00	High
4.01 - 10	Medium
> 10	Low

Table 5-4. Weighted overlay variables, ranking, and weights for the model

Variable	Degrees		Value	Weight
Slope	0- 2		5	20
	2.001 - 5		4	
	5.001 -10		3	
	10.001 - 20		2	
	20 +		1	
Aspect	130-175	W, SW, S, SE	2	5
	176-360 0-129	E, NE, N, NW	1	
	Meters			
Distance from streams	100		4	15
	500		5	
	1000		3	
	2000		2	
	3000		1	
Distance from lakes	100		4	15
	500		5	
	1000		3	
	2000		2	
	3000		1	
Distance from coast and sinuosity	H	100	5	12
	H	500	4	
	H	1000	3	
	H	2000	2	
	H	3000	1	
	M	100	5	8
	M	500	4	
	M	1000	3	
	M	2000	2	
	M	3000	1	
	L	100	5	3
	L	500	4	
	L	1000	3	
	L	2000	2	
	L	3000	1	
Distance from tributary		100	5	8
		500	4	
		1000	3	
		2000	2	
		3000	1	
Distance from archaeology sites		100	3	14
		500	4	
		1000	2	
		5000	1	

were used as limiting factors to determine if each area could have been habitable from 11,000 to 16,000 cal BP. Suites of topologic and hydrologic analyses were used to predict where streams and lakes might have existed prior to inundation by the ocean. All of these data were gathered into an ArcGIS database and used to create the stage two products.

5.3.2.1. Water

Streams or drainage paths were calculated from the one-meter resolution DEM using an improved, highly efficient least-cost-path search (Metz, Mitasova, and Harmon 2011) in GRASS GIS 7 (GRASS Development Team 2015). Flow accumulation was calculated using a constant value for precipitation minus evapotranspiration of 4×10^{-5} mm/s, characteristic of the region based on the TraCE-21K palaeoclimate model (He 2011; Liu et al. 2009). A 0.1 m^3/s discharge threshold was used to define streams. This value, based on records from gauging stations in southeast Alaska, was used to define the headwaters of streams using CUASHI Hydrodesktop (D. P. Ames et al. 2012).

Possible lake locations were generated in a two-step process. First, depressions were identified in ArcGIS 10 with the basin fill algorithm (Spatial Analyst Toolbox). This algorithm typically is used to fill pits caused by data errors in the DEM but also fills enclosed depressions that may have been lake basins. Hence, what are referred to as lakes are actually depressions in the landscape that could have been lakes, marshes, peat bogs, or simply depressions. The second step removed small areas that were not feasible as lakes from the basin-fill output. An area of 200 m^2 was selected as the minimum value. This was an arbitrary value to remove small errors in the file outputted by basin fill.

For each time slice, the stream file was clipped to the palaeo-shorelines from the original stream file. The shorelines were defined as the zero contour of the DEM for each time slice. Stream files had the lakes polygon locations removed so that the streamline file would intersect the polygon lake file and create a tributary at the lake-stream interface (figure 5-7). Tributaries were calculated as the intersection of streams with other streams, lakes, and the coastline.

5.3.3. Stage Two – Model Outputs

The data collected for stage one were analysed and converted into rasters, with buffers, to form intermediate products. These rasters were then weighted together to form a series of final products.

5.3.3.1. Intermediate Products

Each of the intermediate variables is a raster file at two-meter resolution. These include slope, aspect, distance from palaeo-streams, distance from palaeo-lakes, distance from tributary junctions, distance from archaeological sites, distance from palaeo-coast, and coastal sinuosity. All of the variables, except distance from archaeological sites, were derived from the one-meter resolution DEM generated for this project. All of the 'distance from' variables use multiple buffers to delineate the ranked value that was used in the final products.

The GIS Model

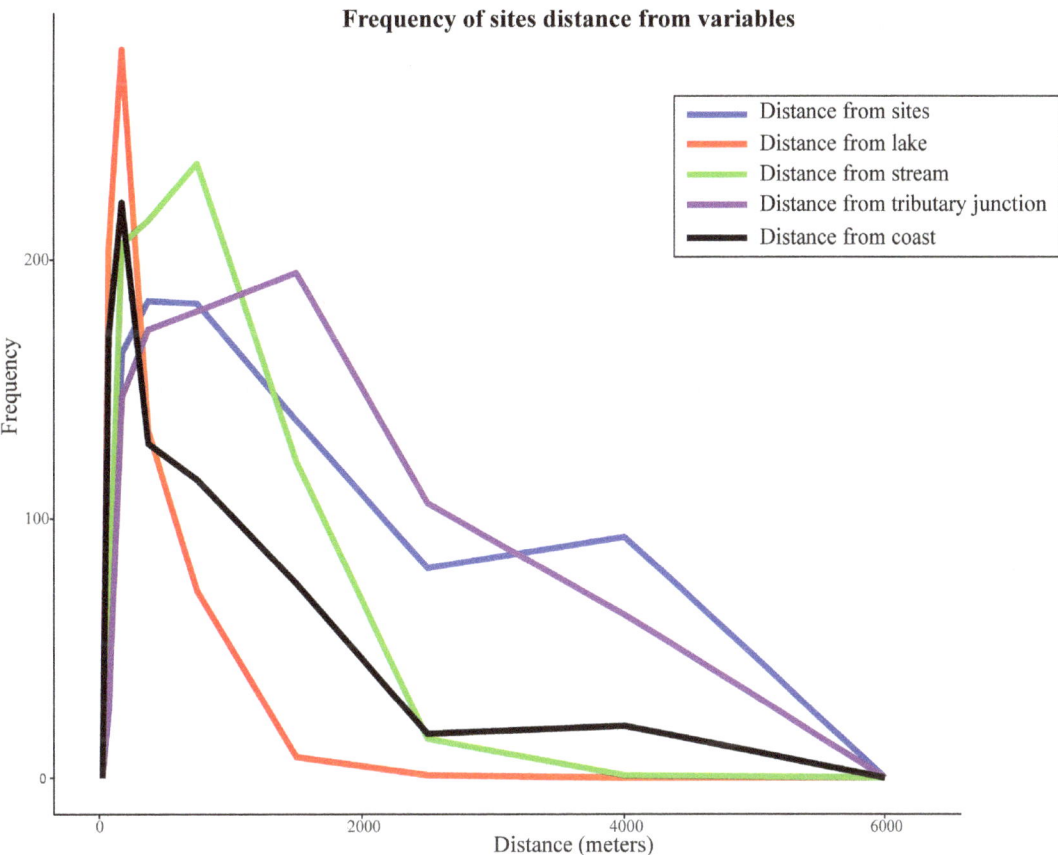

Figure 5-10. Histogram of distance to features (lakes, rivers, tributary junctions, coastline, and other sites) from known archaeological site location.

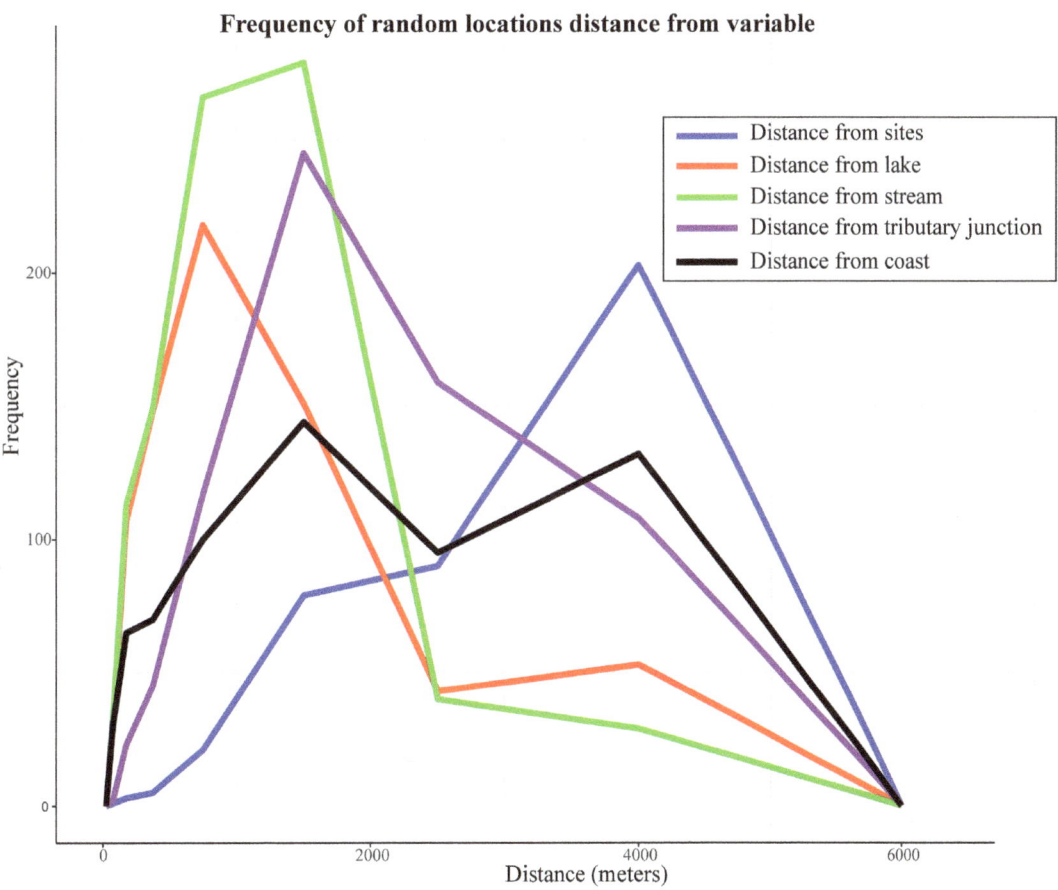

Figure 5-11. Histogram of distance to features (lakes, rivers, tributary junctions, coastline, and other sites) from random locations.

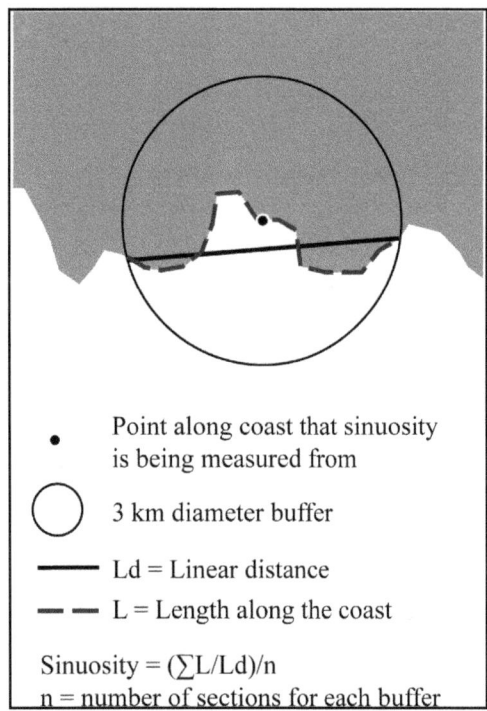

Figure 5-12. Sinuosity is the average of the length along the coastline divided by the linear length of the line segment within a three-kilometre buffer diameter around each point along the coast (Monteleone 2016).

Slope and aspect were generated using ArcGIS 10.0's Spatial Analyst at a resolution of two-meter. This was done for each temporal interval. For each time slice, slope or aspect was reclassified based on the ranked values in table 5-4. These values were developed based on Maschner (1992) and from statistical analysis of archaeological site locations performed specifically for this project. For archaeological sites within the study region, the mean slope is 7.3 degrees and a median slope of five degrees. The mean aspect is 155.3 degrees and the median aspect is 166 degrees (table 5-2). Random locations have a mean aspect of 160 degrees, a median aspect of 158 degrees, a mean slope of 15 degrees, and a median slope of 12 degrees. Because the means and medians appeared similar, Welsh two-sample t-tests (table 5-2) were run in R 3.4.4. The null hypothesis is that the means of the sites and random locations are equal or the same. The slope has a p-value of less than zero (using a 95 per cent confidence interval) and thus the null hypothesis that the mean for site location and random locations are the same can be rejected. This suggests that people chose to live or use the landscape in such a way to create sites, on flat areas. Aspect has a p-value of 0.47. Thus, the null hypothesis that the mean aspect for site location and random locations are the same cannot be rejected (at the 95 per cent confidence interval or alpha of 0.05). Aspect has been weighted low in the models because of the random nature of this variable. This is in contrast to Maschner (1992) who found a southern aspect to be an important variable in site location utilising regression analyses. Because the aspect is the statistically the same between known sites and random location, no inference can be gleamed from the aspect. Slope and aspect were classified for the weighted overlay process (figure 5-8 and figure 5-9). The highest weights were assigned to the lowest slope values (table 5-4).

Using the concept of home range, distance from known archaeological sites is used as a variable to increase the weight for areas known to contain sites. Distance from known archaeological sites was buffered and ranked at distances of 100, 500, 1000, and 5000 m (table 5-4). The highest weight was given to the 500 m bin as the median site distance was 698 m and the mean was 1520.9 m (table 5-2). Twenty-five per cent of the sites in the study area are within 289 m of another site (figure 5-10) and 50 per cent of the sites are within 697 m. Random site locations are usually more than 2000 m away from known sites (figure 5-11). The average distance between the random points (from each other) is 9646 m and the median distance is 6203 m. This means that either people chose to live in the same environment as other people at different times, i.e. around the same site locations, or they chose to live and use the land nearby. This is supported by the Welsh two-sample t-test where the null hypothesis that the means of site distances from other sites is the same as the distance from sites of random points can be rejected (table 5-2).

Streams and lakes were generated for each time slice in stage one. Archaeological site potential around water was ranked from one (lowest) to five (highest) in consecutive buffer rings at 100, 500, 1000, 2000, and 3000 m (table 5-4). The 500 m buffer was ranked higher than the 100 m buffer for lakes and streams based on statistical analysis of the distance from archaeological sites. New analysis using GRASS GIS and R have indicated that in the next iteration of the model, stream and tributary distances should have 500 m buffer ranked higher, not lakes (table 5-2 and figure 5-10). The sites that are further from water could be petroglyphs, pictographs, specific resource patches, or other types of sites such as caves or lithic sources that do not necessarily require water resources. As water sources are depicted as a line with little thickness in GIS, the flood plain and actual size of the water source is unclear. This likely increased the weight of distance from water sources to almost 500 m from known sites, on average. People usually live 'near' water, not necessarily on top of it.

The steps to convert the archaeological sites, streams, lakes, and tributary junctions into raster files were the same. The multi-buffer tool would not work. Hence, each buffer distance was created as a separate file and finally combined together. First, the buffer rings were run and then converted to a raster. The raster was then converted into an integer raster from a floating-point raster. Then the integer raster was reclassified so that the value of the raster was the rank applied to that buffer for that file type (stream, lake, etc.) and 'NO DATA' was zero. The rasters where then added together, two at a time, until all the buffer-rasters are combined into a single output.

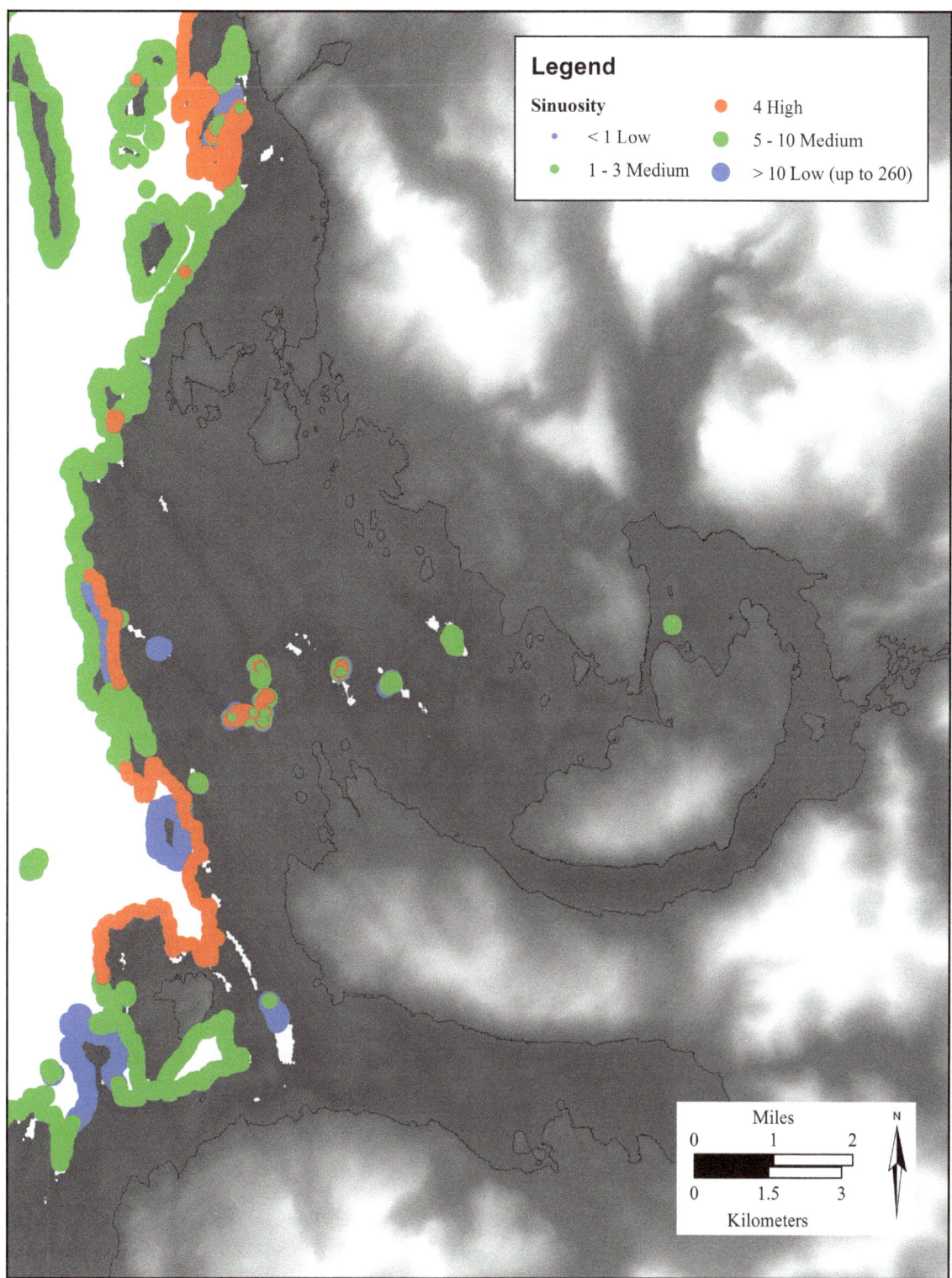

Figure 5-13. Coastal sinuosity of Shakan Bay Area at 13,000 cal BP including high, medium, and low rank for archaeological sites based on statistical analysis.

5.3.3.2. Coasts and Coastal Sinuosity

Coastlines are perhaps the most important resource procurement areas for the NWC. In historic and proto-historic times, ethnographic accounts indicate that the largest native settlements were located in sheltered bays or harbours chosen for a good canoe-landing beach (de Laguna 1960, 30; Meares 1790, 109). For the winter villages containing permanent houses, 'the people prized a view of the more open water across which the canoes of their friends or their enemies might be seen approaching' (de Laguna 1960, 30). The summer fishing camps could be located further up bays at salmon streams (de Laguna 1960; Moss 1992, 7). Therefore, it is not simply the proximity to the coast, but the shape of coast and the character of the offshore environment that is important. Mackie and

Uncovering Submerged Landscapes

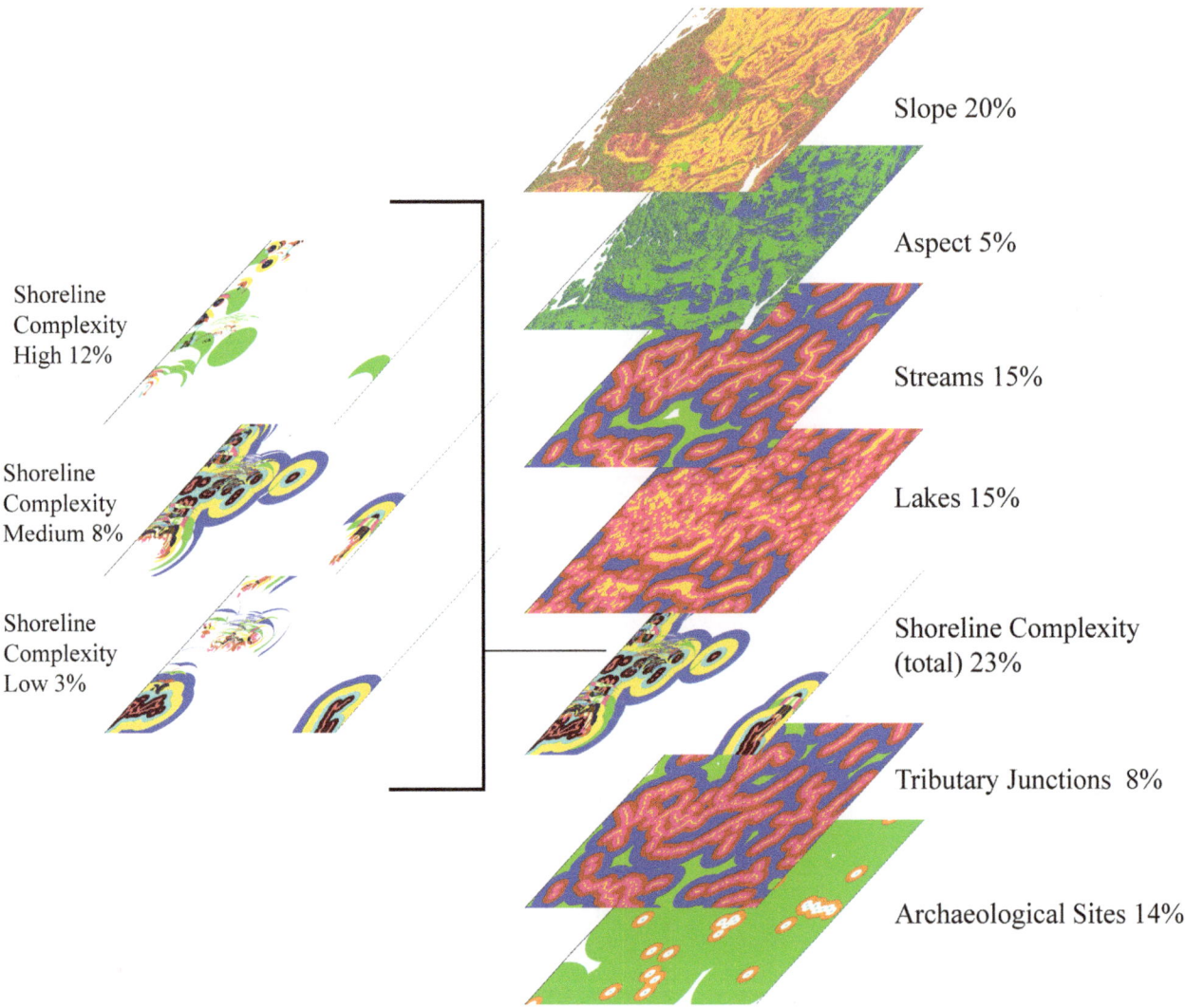

Figure 5-14. Stacked image of rasters used in final weighted model (13,000 cal BP).

Sumpter (2005) utilise shoreline intricacy (like sinuosity) to analyse site distribution patterns on Haida Gwaii.

The coastline was reconstructed at 500 cal BP intervals based on the sea level reconstruction using a contour line at zero m from the DEM. This contour line was assumed to approximate the palaeo-coastline. The coastal sinuosity python script was then applied to the palaeo-coastlines (Monteleone 2013). The mean distance of recorded archaeological sites from the coast is 584 m (table 5-2). However, 50 per cent of the sites are less than 166 m from the coast and 75 per cent are less than 528 m from the shore. Like freshwater sources, people do not choose to live on the water but next to it, hence within 500 m is a reasonable distance to walk to and from the beach. It also provides safety from larger storm surges.

This analysis measured coastal sinuosity using adjacent three km diameter circles along the reconstructed coast using a python 2.6 script (figure 5-12) (Monteleone 2013). The shoreline was clipped to the three km (diameter) buffer at each point along the coast (with a maximum distance of 25 m between points). For each clipped shoreline length, the distance along the coast was calculated. The first and last points for each length of shoreline were used to calculate the linear or euclidean distance using Pythagorean Theorem. The sinuosity of each length of shoreline was then calculated as the distance along the coast divided by the euclidean distance. Finally, the mean value for all the lengths of shoreline within the three km buffer was calculated. The mean sinuosity value was then used for each point along the coast. The values range from linear (zero) to sinuous (values greater than 10). The mean sinuosity value for the modern coast is 4.631. Figure 5-13 shows the coastal sinuosity for Shakan Bay at 13,000 cal BP.

A proximity analysis was run to determine the distance between the location of the recorded archaeological sites and coastal sinuosity values. The mean coastal sinuosity near archaeological sites is 4.7. When compared to the random locations using a Welsh two-sample t-test, the null hypothesis that the mean for the sites is the same for the random locations can be rejected with a p-value of 0.01 or at 99 per cent confidence (table 5-2). This suggests that people may have selected less complex coastlines for

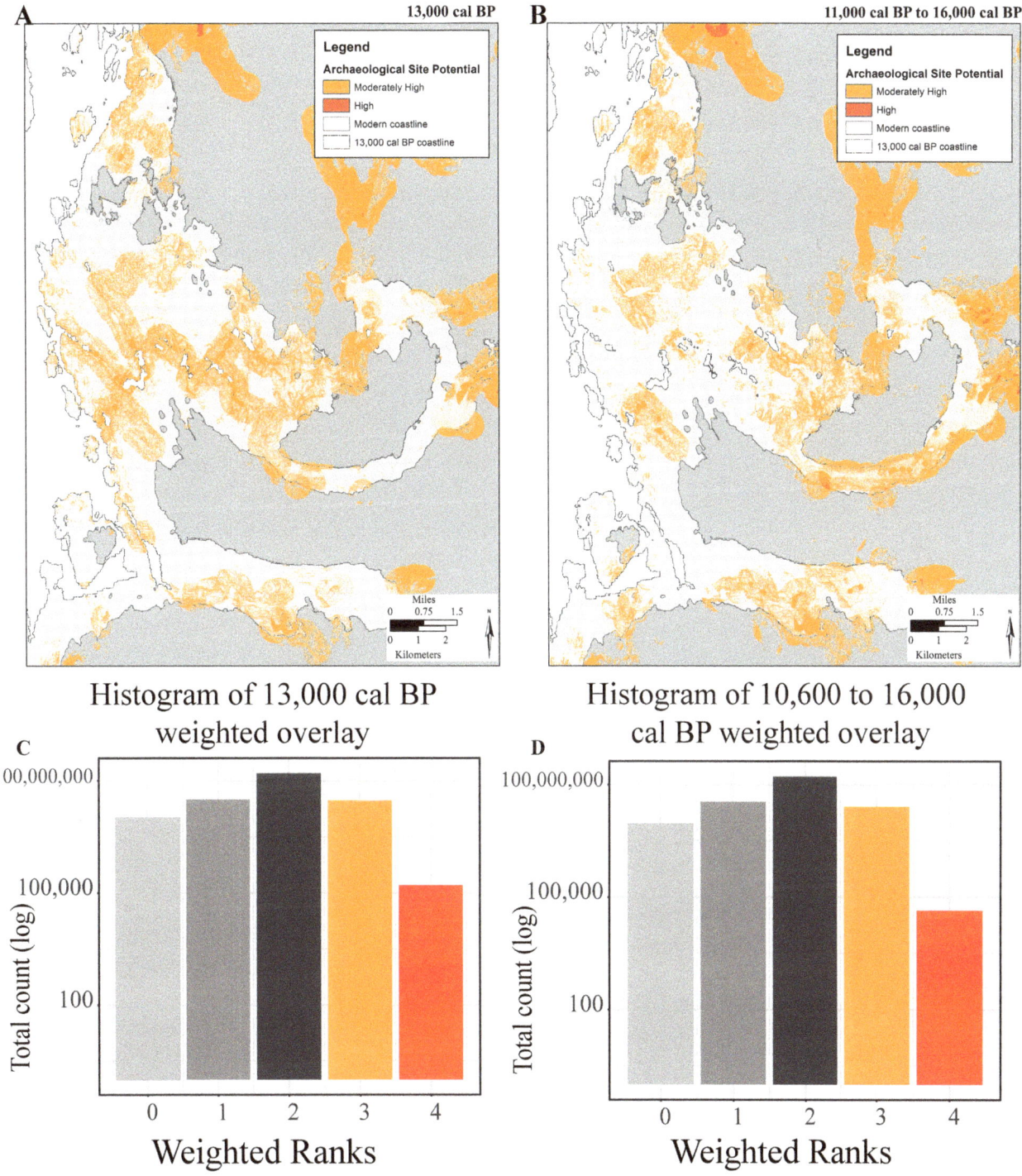

Figure 5-15. Model results. A) Map of Shakan Bay at 13,000 cal BP as an example of a time slice of the final model. B) Map of Shakan Bay with the model results showing the final high potential model with model values from 11,000 to 16,000 cal BP combined. C) Histogram of 13,000 cal BP model based on values. D) Histogram of final model results. The histograms show the full values of the weighted overlay. High potential is value four and moderately high archaeological potential is value three.

land-use and subsistence purposes for the last 5800 years (Monteleone 2016). The coastal sinuosity values for the region were classified into high, medium, and low (table 5-3) based on the proximity analysis from archaeological sites to coastal sinuosity. Distance from the palaeo-shoreline was combined with the coastal sinuosity variable to form a single ranked variable within the model. The high, medium, and low ranks were then buffered in 100, 500, 1000, and 2000 m groups (table 5-4).

5.3.3.3. Final Products

For each time slice, nine raster files (figure 5-14) were combined using ArcGIS Spatial Analyst's Weighted Overlay. The final weights of the model ranged from zero to four. The high potential was assessed based on the distribution of results and determined to be values three and four (figure 5-15). Value three was moderately high archaeological potential and value four is high potential.

The model went through many iterations and attempts to evaluate a reasonable model (Monteleone 2013).

After the model was generated for each of the eleven time-slices, they were combined using ArcGIS 'mosaic to new raster' using the maximum value from each raster. This produced a final model that covered the entire study area from 11,000 to 16,000 cal BP, referred to as the 'deep' model. The resultant ranks ranged from zero to four (figure 5-15). Moderately high archaeological potential and high potential was determined to be the same as the individual time slices, values three and four respectively. This final model accurately incorporates the coasts, lakes, rivers, and archaeological sites in high potential areas.

5.4. Summary

As the research progresses, the iterative process will incorporate new information from analysis, new sites, refined methods, and hopefully, new surveys. This iterative process will continue to refine the model and guide further underwater research in Shakan Bay, which may be applicable to other regions as well.

6

Analysis of the Model

6.1. Introduction

Predictive models need to be tested using a variety of methods to ensure their success. This model has been tested statistically and in the field (chapter seven). Statistical tests include cross-validation using Kvamme's gain, high/low clustering or Getis-Ord General-G, and spatial autocorrelation using Global Moran's I. Statistics were calculated for four models: [1] study region current or modern coastline, [2] study region deep (10,500 to 16,000 cal BP), [3] study area 11,000 cal BP, and [4] study area deep (11,000 to 16,000 cal BP). Random points were used in the statistical test to indicate if the archaeological site statistics were caused by random chance or not.

6.2. Kvamme's Gain

This statistic specifically tests the predictability of the model. The result is a value between negative one and one, presented as per cents. It is the ratio of the number of sites (new, old, or random points) within the area to the per cent of the predicted area from the total area. The formula is gain = 1 − [(per cent of total area covered by model)/ (per cent of total sites within the model area)] (Mink II, Stokes, and Pollack 2006, 215). If the results are negative or less than 50 per cent, the model is not predictive. If it is negatively predictive (result below zero), the areas that are not high potential could and likely have archaeological sites. If the value is between zero and 50 per cent, it means that it is close to equally likely that archaeological sites are in the high potential area or the low potential area. When the gain value is over 50 per cent, it means that more than 50 per cent of the archaeological sites are likely within the high potential areas (Kvamme 1988a; Mink II, Stokes, and Pollack 2006). Table 6-1 shows Kavamme's gain for the four models analysed for the three location groupings.

When comparing within the study region the known archaeological sites file includes 909 sites, there are 22 new sites (table 5-1), and 1000 random points were used. The known archaeological sites are an input into the model, at a weight of 14 per cent. The new archaeological sites are not included in the model. When comparing within the study area the known archaeological site file includes 20 sites, there are 22 new sites (table 5-1), and 50 random points were used. Within the study area, there are so few sites, that none of the known or new sites are in high potential areas. There is no indication that archaeological sites are randomly distributed; these are utilised for comparative purposes only.

Before reviewing the gain statistics, it is interesting to note the per cent of the total area covered by moderately high and high archaeological potential. The area of the study region is over 45,000 km^2; however, the areas of the modern study region and study region deep models is less, 39,069 km^2 and 42,5182 km^2 respectively. The smaller areas are due to 'NO DATA' areas around the outside of the models. The modern study region model has only five per cent of the total area that is moderately high archaeology potential and only 0.02 per cent of the total area that is high archaeological potential. The study region deep model has 14 per cent of its total area as moderately high archaeological potential and 0.07 per cent as high archaeological potential. The study area is just over 1100 km^2; however, the study area 11,000 cal BP model is 658

Table 6-1. Kvamme's Gain statistics

Model		Data Set	Gain Statistic	Predictive Utility (gain)
Study Region (10m model) - 10,500 - 16,000 cal BP				
moderately high potential		Known sites	84%	Positive
		New sites	81%	Positive
		Random locations	-0%	None
high potential		Known sites	99%	Positive
		New sites	null	
		Random locations	66%	Positive
all high potential		Known sites	85%	Positive
		New sites	81%	Positive
		Random locations	-0%	None
Study Region (10m model) - modern				
moderately high potential		Known sites	92%	Positive
		New sites	90%	Positive
		Random locations	-0%	None
high potential		Known sites	100%	Positive
		New sites	null	
		Random locations	79%	Positive
all high potential		Known sites	92%	Positive
		New sites	90%	Positive
		Random locations	-7%	negative
Shakan Bay (2m model) 11,000 - 16,000 cal BP				
moderately high potential		Known sites	84%	Positive
		New sites	76%	Positive
		Random locations	8%	None
Shakan Bay (2m model) 11,000 cal BP				
moderately high potential		Known sites	81%	Positive
		New sites	78%	Positive
		Random locations	-21%	Negative

km² and the study area deep model is 913 km². Again, some of the missing areas are accounted for by 'NO DATA' areas. The study area 11,000 cal BP model has 12 per cent of the model area as moderately high archaeological potential and 0.08 per cent as high archaeological potential. The study area deep model has 13 per cent of the total area as moderately high archaeological potential and 0.07 per cent as high archaeological potential. The sustainably lower areas of the modern study region are reflected in the gain statistics.

The study region deep model, 10-meter resolution combining models from 10,500 to 16,000 cal BP has some positive predictive utility, but that is negated by the positive predictive gain of the random locations. Despite the moderately high archaeological potential predictive values of 84 per cent and 81 per cent for known and new archaeological sites and 99 per cent predictive utility for known sites in high archaeological potential areas, there is 66 per cent predictive utility for random location in the high archaeological potential areas. There are two random points that land in the high archaeological potential area; statistically this is expected as it was discussed in chapter two in relation to non-sites. Due to the very small area covered by the high archaeological potential area, this is highly improbable and detracts from the usefulness of the study region deep model.

The modern model for the study region follows a similar pattern to the study region deep model with very high predictive utility based on known and new archaeological sites, but also a high gain value for random locations within high potential areas. This is a single random point out of 1000 but still derives a gain statistic of 79 per cent. The known and new archaeological sites have over 90 per cent positive predictive gain, except there are no new sites within the high archaeological potential areas. Generally, the modern study region model and the study region deep model are both positively predictive but need improvement. This was one of the many reasons for developing the two-meter resolution models for the study area.

The study area models at two-meter resolution produced considerably improved predictive utility with moderately high positive grain values for the known and new sites and no gain or negative gain values for the random locations. However, there are no sites located in high potential areas, only in moderately high potential archaeological areas. A similar pattern emerges in chapter eight when the known pre-10,000 cal BP northern NWC sites. The lack of sites in high potential areas is likely due to the extremely small area of high potential, 0.07 per cent, and 0.08 per cent of the models in the study area (1126 km²) or approximately 620 m² and 540 m². The positive predictive utility of these smaller models is slightly lower than the study region models, but the gain values are still higher than 76 per cent for new archaeology sites and 81 per cent for known archaeological sites. The study area deep model, which includes the 11,000 to 16,000 cal BP models, produces an 84 per cent predictive gain for known archaeological sites, 76 per cent for new archaeological sites, and eight per cent for the random locations. As a 'current' final model, these values are encouraging. The 11,000 cal BP model is very close to the modern coastline, but only has an 81 per cent positive gain for known sites, 78 per cent for new sites, and -21 per cent for random locations. When trying to locate sites of this sea level, this model is highly productive.

Overall, the gain statistics for the four models are very promising and with additional survey and testing, archaeological sites are expected to be located. The chance of false positives from the study region models may be great in the high archaeological potential areas, but the focus moving forward is on the study area and the two-meter resolution models that do not have this issue.

6.3. Spatial Statistics

Regional or landscape scale spatial analysis has a long history in archaeology (Willey 1953; Kohler and Parker 1986). Archaeological theories often focus on the spatial distribution of materials within archaeological sites or intra-site statistics (Nobles 2013; Ahler 1989; Nance and Nance 1986; McCoy and Ladefoged 2009; Wheatley and Gillings 2002). There the focus is on between site or inter-site statistics; what is tested here is the model locations using similar assumptions as inter-site analyses. Based on archaeological theory of resource procurement (Binford 1980), archaeological sites should demonstrate high clustering or autocorrelation. This pattern is found when analysing the four models created here. 'Perhaps in building predictive models we are too ready to make the assumption that only a complex multivariate model can adequately account for human location behaviour, when in fact a few (proxy?) variables, observed in the highly correlated database that is our environment, may be sufficient for forming locational decisions' (Kohler and Parker 1986, 433).

A cluster is 'a number of similar things that occur together' (MerriamWebster 2018, 'cluster'). Something is correlated when a relation exists between phenomena that tend to vary, be associated, or occur together in a way not expected on the basis of change alone (MerriamWebster 2018, 'correlation'). Autocorrelation is 'the correlation between paired values of a function of a mathematical or statistical variable taken at usually constant intervals that indicates the degree of periodicity of the function' (MerriamWebster 2018, 'autocorrelation').

6.3.1. Getis-Ord General-G

The General G statistic tests the amount of clustering in both high and low values. It is different from Moran's I because it incorporates the high and low values. The null hypothesis is that there is no spatial clustering in the data. The null hypothesis can be rejected when the p-value returned is small and statistically significant at a confidence interval of 95 per cent or greater. If the null hypothesis is rejected, then the sign of the z-score becomes important. The z-score is a standard score, which indicates how many standard deviations, and observation is above or below the mean. If the z-score value is positive, the observed General G value is larger than expected. This indicates that high values for the model are clustered. If the z-score value is negative, the observed General G value is smaller than expected, indicating that low values are clustered (ESRI 2012a). Though the null hypothesis is for no correlation, the expected outcome is high correlation due to the clustered nature of archaeological sites and resources.

Getis-Ord General-G is a tool in the Spatial Statistic Toolbox for ESRI's ArcGIS10. It requires a vector input. The models were created as raster files. Before converting the model, with values ranging from zero to four, to polygons, values from zero to two were removed in order to simplify the polygons and decrease the size of the data. The only input for this tool is the polygon of the moderately high and high potential areas.

Table 6-2 has results of the Getis-Ord General-G statistic for each model using inverse-distance. All results are statistically significant and the null hypothesis can be rejected for all four models. Essentially, there is high clustering in all four models. The z-scores are very high for the study region models and high for the study area models; a similar pattern is found with the Moran's I statistic. The distance threshold for the models decreases from the high of almost 6000 m for the study region deep to 458 m for the study area 11,000 cal BP model. The study region deep model's high distance threshold indicating there are few nearby neighbours or the high potential areas are spread out but clustered, likely because of the limited high potential (value four) in the model.

6.3.2. Global Moran's I

Global Moran's I tests spatial autocorrelation. The null hypothesis is that there is no spatial correlation in the data. The Global Moran's I, or spatial autocorrelation tool, measures spatial autocorrelation based on feature locations and feature values simultaneously. Given a set of features and an associated attribute (value three or four), the tool evaluates whether the pattern expressed is clustered, dispersed, or random. It computes the mean and variance for the model value. Then, for each value, it subtracts the mean, creating a deviation from the mean. Deviation values for all neighbouring features (features within the specified distance band, for example) are multiplied together to create a cross-product (ESRI 2012b).

Moran's I is a tool in the Spatial Statistic Toolbox in ESRI's ArcGIS10. The same polygon features created for Gets-Ord's General-G were utilised. Table 6-2 has the results of the Global Moran's I run using inverse-distance. All of the models are highly autocorrelated, but interestingly, the inverse distance threshold values show a similarity between the Getis-Ord General-G statistic and the Global Moran's I statistic. The value is actually identical for the modern study region model and is only a few tens-of-meters different for the study region deep and study area 11,000 cal BP. For the study area deep model, the values are almost half of the threshold value for the Getis-Ord General-G.

Global Moran's I was also calculated at 250, 500, 1000, 1500, 2000, and 3000 m bands or radius (table 6-3). The results indicate that the models are highly autocorrelated. There are no neighbours for a number of the models at lower distance bands. The tool parameters clearly state that all features must have a neighbour to be valid (ESRI 2012b). Getis and Ord (1992) do not include this as a requirement for I statistics. This error occurs for all bands smaller than 1500 m for the study region deep model and study area model, for all band smaller than 2000 m for the modern study region, and for the 250 m radius for the study area 11,000 cal BP model. This means that the results are not very tightly clustered given the lack of neighbours within these relatively close distances.

Table 6-2. Getis-Ord General-G and Global Moran's I statistics using inverse-distance

		Study Region Deep	Study Region Modern	Study Area Deep	Study Area 11,000 cal BP
Getis-Ord General G	Observed General G	0.000	0.000	0.000	0.000
	Expected General G	0.000	0.000	0.000	0.000
	Variance	0.000	0.000	0.000	0.000
	Z-score	45.997	37.339	13.916	5.020
	p-value	0.000	0.000	0.000	0.000
	Distance threshold	5984 m	3640 m	1208 m	458 m
Global Moran's I	Moran's Index:	0.185	0.328	0.664	0.730
	Expected Index:	-0.000	-0.000	-0.000	-0.000
	Variance:	0.000	0.000	0.000	0.000
	z-score:	1992.912	1471.494	19416.755	15298.541
	p-value:	0.000	0.000	0.000	0.000
	Distance threshold	5396 m	3640 m	678 m	423 m

Table 6-3. Global Moran's I results using fixed distance bands

		Nearest Neighbor Radius (meters)					
		250	500	1000	1500	2000	**3000**
Study Region - Deep (10,500 - 16,000 cal BP)	Moran's Index:	0.671475	0.527092	0.314336	0.195419	0.135984	0.083321
	Expected Index:	-0.000001	-0.000001	-0.000001	-0.000001	-0.000001	-0.000001
	Variance:	0.000000	0.000000	0.000000	0.000000	0.000000	0.000000
	z-score:	1532.221050	2034.925966	2092.007197	1807.976948	1597.086898	1382.976612
	p-value:	0.000000	0.000000	0.000000	0.000000	0.000000	0.000000
	no neighbors	4092	286	27	8	0	0
Study Region - Modern Coast	Moran's Index:	0.741068	0.611480	0.371596	0.241848	0.168449	0.094450
	Expected Index:	-0.000003	-0.000003	-0.000003	-0.000003	-0.000003	-0.000003
	Variance:	0.000000	0.000000	0.000000	0.000000	0.000000	0.000000
	z-score:	1269.843242	1700.647814	1663.539672	1428.874939	1216.664220	918.359416
	p-value:	0.000000	0.000000	0.000000	0.000000	0.000000	0.000000
	no neighbors	2017	193	25	9	3	0
Study Area Deep (11,000 - 16,000 cal BP)	Moran's Index:	0.485152	0.359042	0.198884	0.114747	0.076389	0.056285
	Expected Index:	-0.000009	-0.000009	-0.000009	-0.000009	-0.000009	-0.000009
	Variance:	0.000000	0.000000	0.000000	0.000000	0.000000	0.000000
	z-score:	1486.017416	1742.054729	1527.236510	1149.248329	937.066378	949.622968
	p-value:	0.000000	0.000000	0.000000	0.000000	0.000000	0.000000
	no neighbors	44	2	1	0	0	0
Study Areas 11,000 cal BP	Moran's Index:	0.376813	0.327911	0.217819	0.134199	0.094209	0.075453
	Expected Index:	-0.000014	-0.000014	-0.000001	-0.000014	-0.000014	-0.000014
	Variance:	0.000000	0.000000	0.000000	0.000000	0.000000	0.000000
	z-score:	976.617811	1322.324728	1348.258717	1039.926131	863.867722	923.954708
	p-value:	0.000000	0.000000	0.000000	0.000000	0.000000	0.000000
	no neighbors	33	0	0	0	0	0

Generally, Moran's I statistic ranges from negative one to one. The results for the model analysis from 0.05, for 3000 m radius study area deep, to 0.7, for the 250 m radius modern study region. The lower values are associated with the larger neighbourhoods (3000 m) and the higher values with the smaller neighbourhoods (250 m). The smaller neighbourhoods have locations that do not have neighbours and are thus statistically insignificant values. All of the results are positive and show some clustering. The z-scores for all of the results are very high, over 100, which indicate the results are significant at five per cent alpha. The p-values are all less than zero. This means the results are significant, and therefore, there is less than a one per cent likelihood that the results are random. The null hypothesis can be rejected; the models show autocorrelation.

The results are clustered likely because the environmental variables that they are mirroring are clustered. Though the Moran's I statistic was suggested to be a useful test of a predictive model, these results are more descriptive. They are not directly useful for testing the model results. The Moran's I statistic indicates that the model is clustered, which is expected from freshwater and coastally focused variables. The archaeological sites were not included in some of the earlier iterations of the model that still demonstrated high spatial autocorrelation (Monteleone 2013). Therefore, the archaeological sites are not causing the clustered results.

6.4. Summary

Kvamme's Gain indicates that the study region models are slightly predictive and the study area models are moderately predictive. Because the modelling process is iterative, the goal of future work will be to increase this value, which has already been achieved from the last iteration (Monteleone 2013).

The lack of near-neighbours in the study region model is surprising due to the large areas covered by the moderately high and high potential areas, but is clear from the distance thresholds in table 6-2 and the large number of features with no neighbours in table 6-3 that the data is spread out. Some degree of clustering should be expected in archaeological sites based on the logistical collectors model by Binford

(1980) and the ethnographic overview presented in section 4.5 (Northern NWC Ethnographic Groups). The clustering is also expected because environmental variables that are being modelled are clustered, such as the coast and rivers. This is supported by both the Getis-Orts General-G high/low culturing and the Moran's I statistics.

7

Field Testing the Model

7.1. Introduction

There have been two opportunities to field test the high potential model as part of the Gateway to the Americas I and II projects. The first in June 2010 was a preliminary survey to demonstrate that the survey methods were adequate and appropriate for southeast Alaska (Dixon and Monteleone 2011). A few anomalies were identified in the geophysical survey that helped focus a new NSF proposal for three years of research. The Gateway to the Americas II project received 10 days a year support of ship time for marine survey from 2012 to 2014. Testing was done in other areas within the larger study region during 2013 and 2014 (Monteleone and Dixon 2018). The primary goal for 2012 was collecting data to test the high potential model. The field results presented here are from the 2010 and 2012 surveys.

7.2. June 2010 Survey

The 2010 fieldwork was conducted between 16 and 24 June 2010, aboard the 86' (26.2-meter) *FV Crain*. The survey team consisted of Dr E. James Dixon (PI), the author, and two technicians: Tim Bulman of InDepth Marine and Harry Castle of Aquasphere Surveyors Inc. The ship was crewed by Chris Beaudin, Posi Beaudin, Gus Beaudin, and first mate James Bacon. As this project included other survey areas not covered here, only one day of survey was conducted on 18 June 2018 in Shakan Bay. The seafloor was imaged using a dual beam adjustable frequency Imagenex Model 872 Yellowfin side scan sonar. Trimble XRS DGPS with integrated (internal) coast guard beacon provided real-time satellite corrections and sub-meter accuracy.

Figure 7-1. Map of Shakan Bay side scan survey from June 2010 indicating anomaly locations. The modern land extent is in grey with model the background.

Side scan sonar uses pulses of sound emitted in a conical, or fan shape, directed towards the seafloor. The acoustic reflection of the seafloor is recorded as intensity on either side of the vessel. The Yellowfin can collect 1000 data points per side, there is one transducer per side that is tilted down at 20 degrees. Three different frequencies are available: high, medium, and low. Higher frequencies provided better resolution but cover less area. For the Yellowfin, high frequency is 800 kHz with a beam width of 0.7 degrees by 30 degrees. This means that each time a pulse is emitted it is 0.7 degrees wide and spread over 30 degrees. The number of data points is then spread over the entire beam area per side. Medium frequency is 330 kHz with a beam width of 1.8 degrees by 60 degrees. Low frequency is 260 kHz with a beam width of 2.2 degrees by 75 degrees (Imagenex 2011). The medium setting was determined to be the most effective for survey purposes and provided 125 m of visibility on either side of the vessel. Unfortunately, a calculation error during survey preparation for Shakan Bay resulted in the survey lines not overlapping, as was intended.

The side scan sonar was towed beside the vessel using 'cannonball' weights to keep the line and tow-fish (side scan unit) as vertical as possible and operating at appropriate depths beneath the surface. The exact location of the side scan in relation to the vessel varied with speed, current, length of extended cable, and depth. The *F/V Crain* was unable to maneuver at the slow speeds required to operate the side scan. The helmsman was required regularly to shift in and out of gear while trying to steer a straight path. The results were sinuous, rather than straight, paths with variations in the location of the side scan in relation to the GPS receiver, and thus a loss of accuracy in identifying the location of anomalies. The side scan did not have an attached transceiver to indicate its precise location. Thus, side scan anomalies are several meters from their precise locations.

Shakan Bay is sheltered from the Pacific Ocean by Kuiu Island to the west. Four side scan transects were run from west to east, with half of a transect attempted from east to west. The current was from west to east and made it difficult to keep a straight transect from east to west, partially because of the slow speed required for the survey. Originally, five anomalies were identified (figure 7-1) but subsequently number four could not be identified in the side scan data. Figure 7-1 includes the current predictive model for Shakan Bay, the side scan survey paths, and identified anomalies.

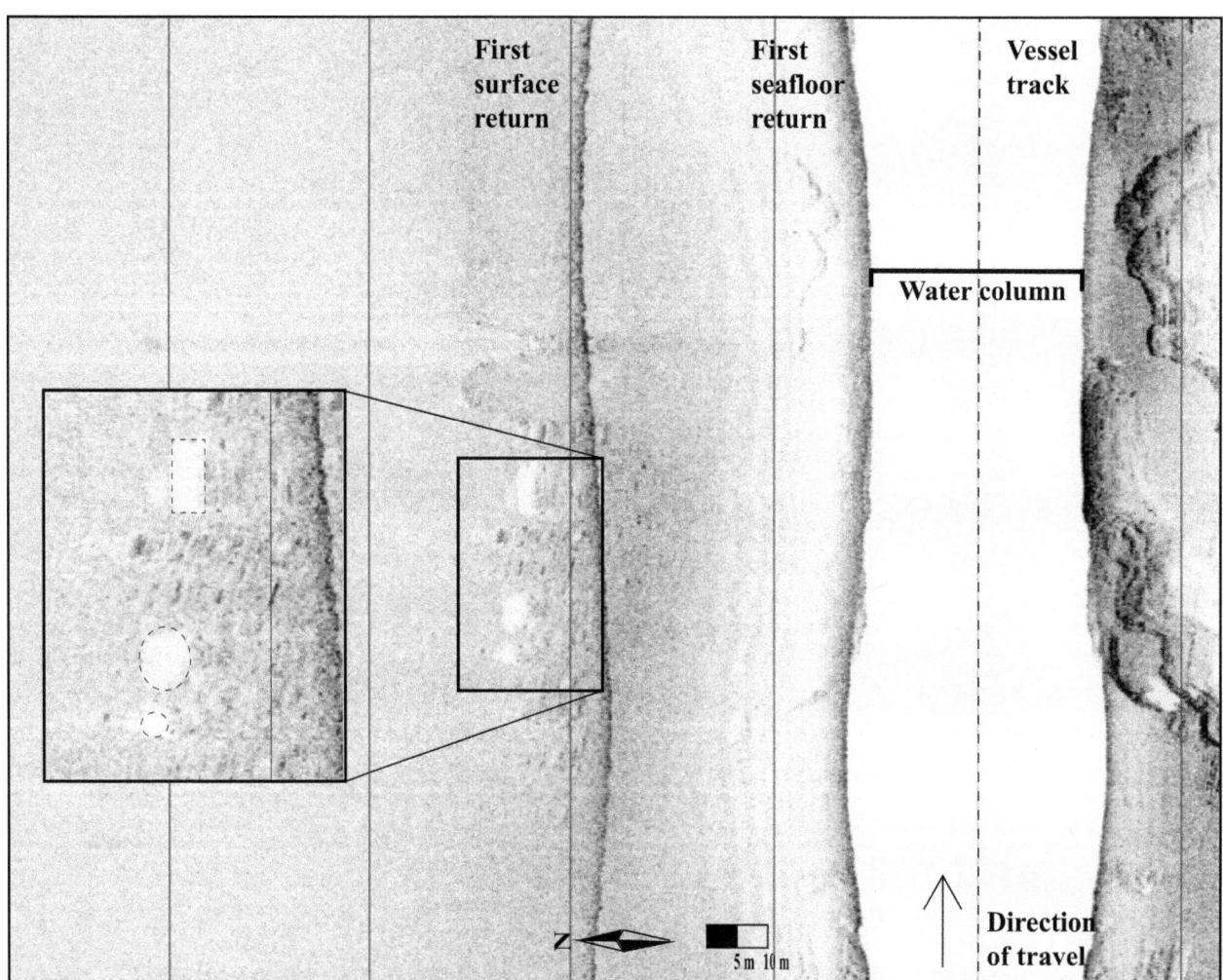

Figure 7-2. Shakan Bay Anomaly one: two circular and one rectangular depression (white space). Inset has depressions outlined in dotted lines.

Shakan Bay is a likely location to identify archaeological sites that are older than 10,600 or even 13,000 cal BP. The majority of the area was sub-aerial prior to sea level rise to a depth of 60 m or 12,500 cal BP. Though the coast extended further seaward in Sumner Strait, palaeogeographic reconstruction suggests that rivers and lakes may have been favourable habitat to early settlers. Sumner Strait contained a large glacial lobe during the LGM, but the lobe is expected to have retreated to the north end of POWI by about 15,000 cal BP (Carrara et al. 2003; Carrara, Ager, and Baichtal 2007; Shugar et al. 2014). This means there was deep-water habitat available for fishing adjacent to Shakan Bay during the late Pleistocene and early Holocene.

7.2.1. Shakan Bay Anomaly One

Anomaly one in Shakan Bay consists of two circular and one rectangular depression (figure 7-2). The shadows are white space on a grey background, using the reverse grey colour table. There are no dark or raised returns between the sonar and the shadows, which indicate they are depressions. The rectangular depression is approximately two meters wide and five meters long. The larger circle has a diameter of approximately three meters. The smaller circle, on the bottom of the image, has a diameter of 2.2 meters. The anomalies are in approximately 60 meters of water. This image is from the port side of the side scan. The sonar sends pulses of sound in all directions and records each return. The surface of the water occurs as a solid or dark line on each side (port and starboard) of the sonar image. The depth of an anomaly is calculated by adding the distances to port or starboard of the anomaly and the distance port or starboard of the surface return.

The circular and rectangular shaped depressions could be cache or storage pits or house depressions similar to those found in southeast Alaska during the historic (PET-00126), proto-historic, and developmental phases (Twelve Mile Midden – PET-00493, Kutsnoi Habitation (Baby Bear Bay 2 – SIT-00353). Twelve Mile Midden was occupied between 664 and 782 cal BP (660±60 and 810±60 ^{14}C years) (Beta-158919 and Beta-158918) and includes a depression measuring approximately 3.2 m long, 1.2 m wide, and 30 cm deep. Kutsnoi Habitation has a depression that is 2.35 by two meter and is 30 cm deep and dates to 1430 cal BP (Beta-56342). Baby Bear Bay 5 site has two oval depressions; one depression is 2.5 by three meters and the other is five meters square. There is also a rectangular depression, which is 2.5 by three meters. The site dates between 940 and 1220 cal BP (Beta-56345, Beta-56346) (Moss 1998; Bower 1994). Depressions are common features through

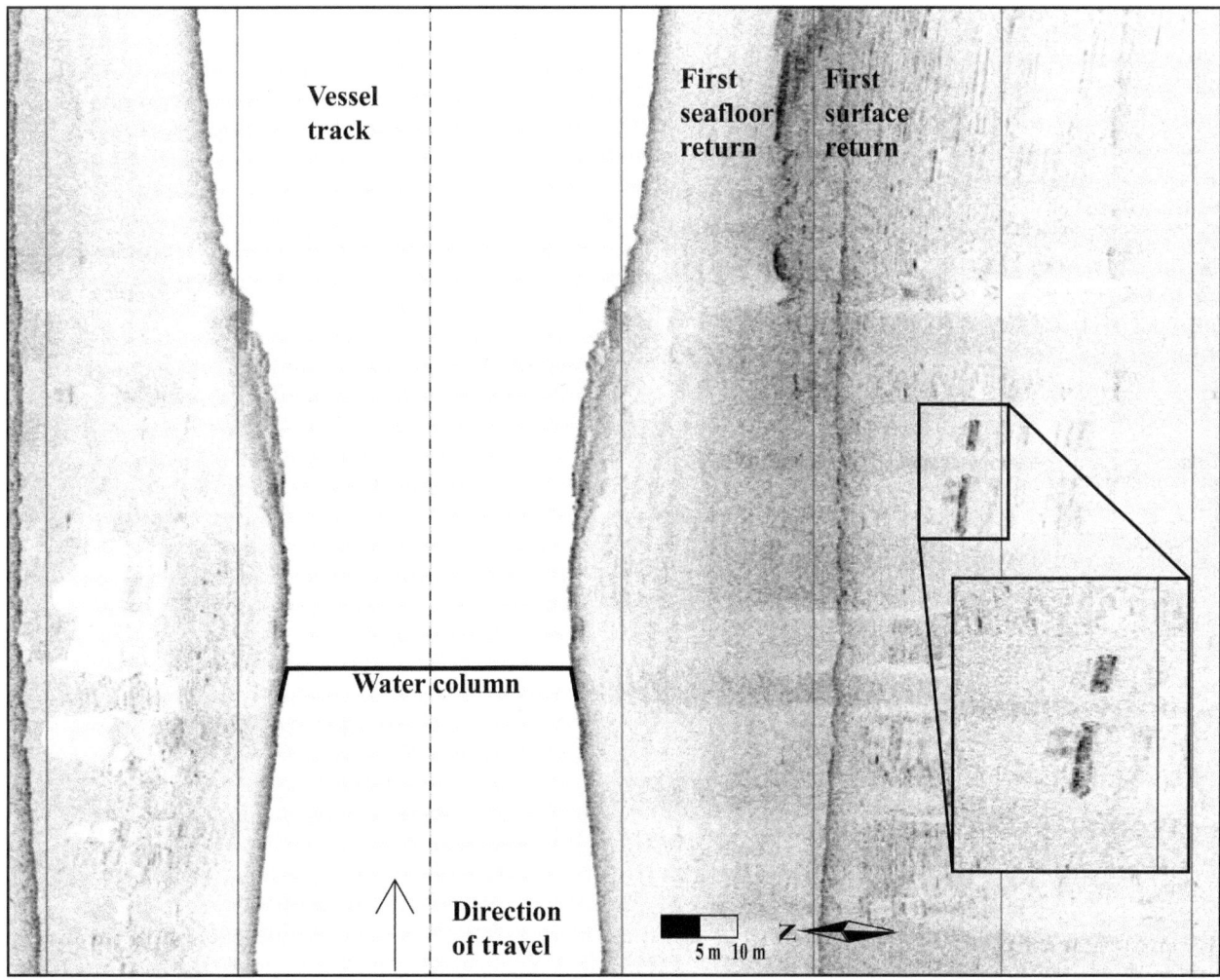

Figure 7-3. Shakan Bay anomaly two: two raised rectangular features (darker grey).

the archaeological record for southeast Alaska. These three depressions could be cultural in origin based on comparison with similar recorded archaeological features.

7.2.2. Shakan Bay Anomaly Two

Anomaly two in Shakan Bay consists of two raised rectangular features (figure 7-3). The top rectangle is approximately one meter wide by 4.5 m long. It is roughly 40 cm high. The lower rectangle is approximately one meter wide and nine meters long. It is 40 cm high. It is unclear how much of the anomaly is below sediment and thus it is difficult to ascertain its true height. The area surrounding this anomaly is mainly flat sandy sea bottom. The anomaly is located at 71 m depth and is located on the starboard side of the sonar. The rectangular features appear to be human-made, but it is unclear what they are. They could be from logging activities in the early 1900s. One suggestion was that it could be a log or cut tree.

7.2.3. Shakan Bay Anomaly Three

Anomaly three in Shakan Bay was interpreted from the side scan records as several stone-semi-circular raised features and two circular depressions (figure 7-4). The semi-circles are between 20 and 40 cm high and 3.5 to four meters wide. There are seven or eight of these semi-circular features near two depressions. The depressions are approximately 2.5 m in diameters. The depth below sea level is 52 m and the anomalies are located on the starboard side of the ship. It is unclear what the two depressions are, but one suggestion is that they are freshwater vents or similar karstic features.

One hypothesis to explain the semi-circular features is that they are intertidal fish weirs. Smith (2011) identified five types of fishing structures in southeast Alaska: baskets (0.8 per cent), stone (45.5 per cent), wood stake (48 per cent), basket and stake (0.3 per cent), and stone and stake (5.4 per cent). The per cent provided are for the number of recorded fishing structures in southeast Alaska from a total of 369. The stone and stone with stakes weirs combine to create the largest category of fishing structures. Similarities can also be drawn between these semi-circular anomalies and several sites in southeast Alaska including CRG-123, CRG-269, CRG-369, and PET-125 (Smith 2011). Figure 7-5 is a reconstruction of the landscape when the potential fishing weir and depressions were exposed as dry land, using QPS Fledermaus. At a depth of 52 m, the region would have been an intertidal estuary. This landscape

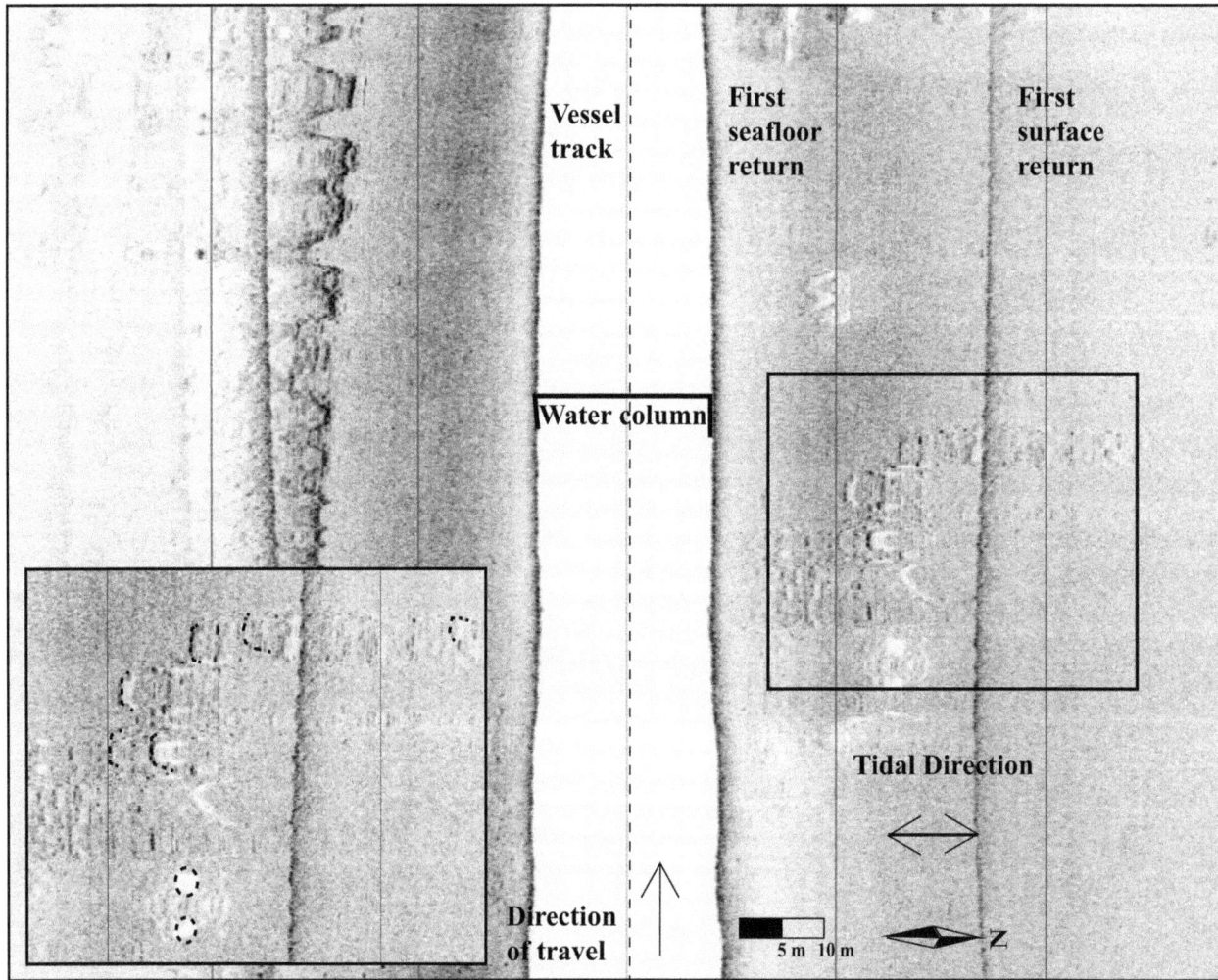

Figure 7-4. Shakan Anomaly three: raised semi-circular features and two depressions. Inset has raised features and circular depressions indicated with dotted lines.

Field Testing the Model

Figure 7-5. Reconstruction of the location of the potential weir and depressions on the landscape (sea level is at 52 m with a one-meter gap for tide).

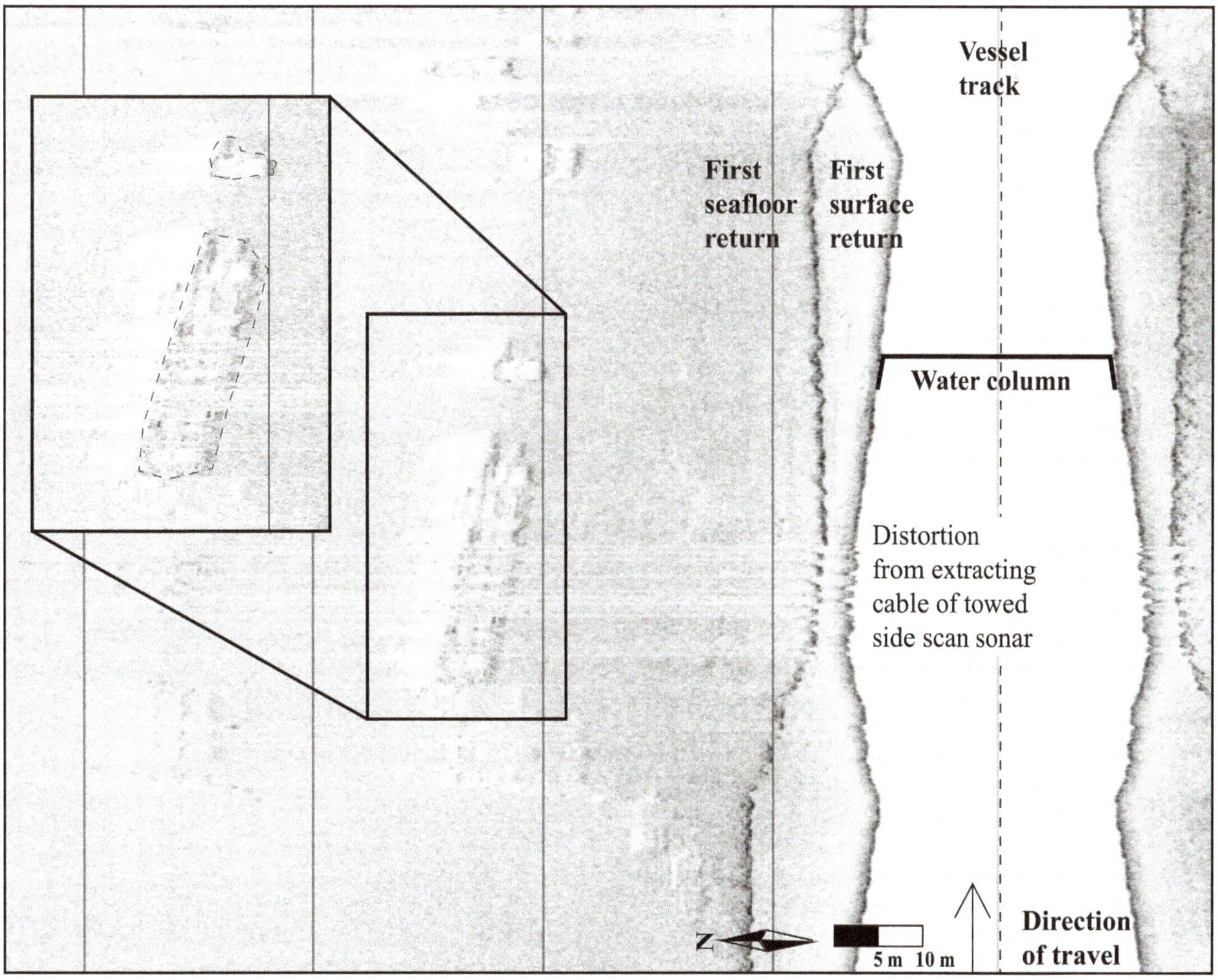

Figure 7-6. Shakan anomaly five: shipwreck (PET-714), likely the Restless lost in 1910. Inset has raised features outline with a dotted line.

would have existed approximately 12,300 cal BP, based on the sea level reconstruction. If the semi-circular features were fish weirs, they would be the oldest fish weirs in North America. Ancient fish weirs are found throughout southeast Alaska and the oldest weirs are approximately 6000 cal BP (Smith 2011). Older structures likely exist but have not yet been identified. This location was investigated during the May 2012 field season.

7.2.4. Shakan Bay Anomaly Five - Shipwreck

Shakan Bay anomaly five is a shipwreck (Figure 7-6) at approximately 35 m of water. The vessel has the bow or stern detached, so it is in two parts. The main part of the anomaly is approximately six meters wide and 35 m long. The anomaly is one to 1.5 m above the seafloor. Figure 7-6 depicts both the raised anomaly in grey and the shadow in white behind the two pieces of the possible wreck.

The Restless is a Yawl vessel that sank 10 February 1910, between Shipley Bay and Shakan. It was stranded when the anchor line broke. Her last port was Wrangell. The vessel is of little value (financially or regarding cargo) and there are currently no other details regarding the vessel. Anomaly five in Shakan Bay is around 2.7 km (1.7 miles) from the reported location of the Restless. This reported location is a point that was put on a map based on survivor's descriptions of where the vessel sank and consequently it is not precise. The anomaly could be the Restless and the Alaska SHPO office has identified the location as PET-714.

7.3. May 2012 Survey

Fieldwork was conducted from 21 – 31 May 2012 with the 86' *F/V Crain* and the 36' *R/V Antonie*. The *R/V Antonie* is a fibreglass gillnetter. The *F/V Crain* was utilised as a platform for deploying the ROV and van been sampling, screening, and crew accommodations during the survey. The *R/V Antonie* was used for surveying using multibeam sonar and sub-bottom profiling.

The archaeological team consisted of E. James Dixon (PI), Forest Haven, and the author. Forest Haven is Tsimshian from Metlakatla. She was an undergraduate student at the University of Alaska Southeast and was the intern selected by Sealaska Heritage Foundation to participate in the 2012 season. The technicians consisted of Tim Bulman of InDepth Marine, Harry Castle of Aquasphere Surveyors Inc., and Rick Hollar of AUS Diving. The crew of the *F/V Crain* was Chris Beaudin, Conrad Beaudin, and Stacie Murray. Tony D'Aoust is the owner and captain of the *R/V Antonie*.

Two areas were surveyed during the 2012 survey: Shakan Bay and the Gulf of Esquibel. The Gulf of Esquibel data will not be discussed here. An area of 13.8 km² was surveyed in Shakan Bay from the 23 of May through the 26 of May. Two new locations, in addition to the five identified on the side scan data, were tested during this survey, Shakan six and seven. The sub-bottom profiler was used to resurvey the same area as the multibeam data on 26

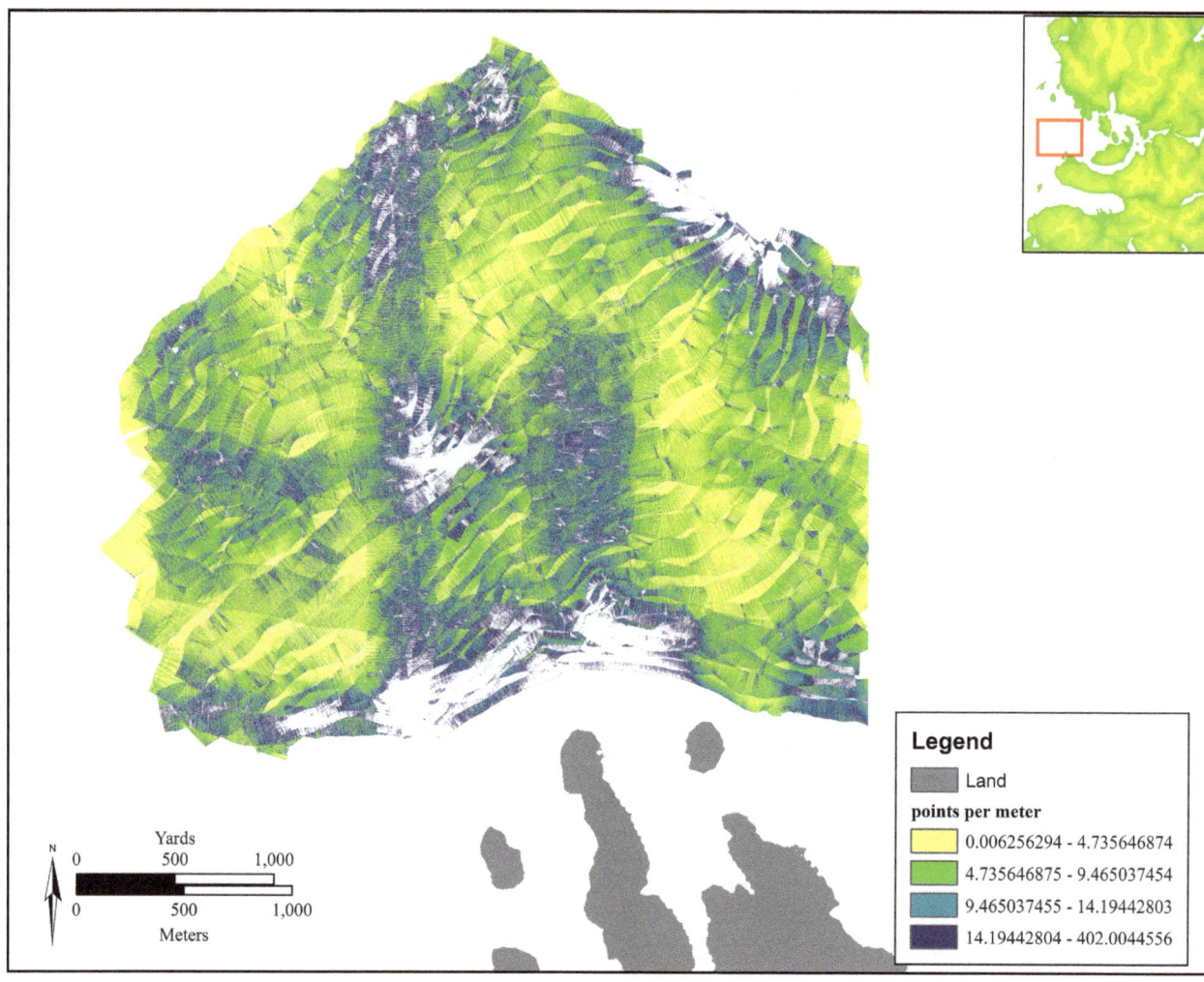

Figure 7-7. Density of multibeam points per 1m². Gaps in multibeam coverage are clearly visible as white space.

and 27 of May. However, it was not correctly calibrated, and as a result, the product is limited or poor quality sub-bottom data for Shakan Bay.

The location of the *R/V* Antonie was recorded by Hemisphere V100 GPS dual antenna systems. This tracked the position and heading providing differential or DGPS for the survey. They were also calibrated to record the pitch, roll, and heave of the vessel. The same equipment was used aboard the *F/V Crain* for the ROV and van veen positions. Sensors were mounted on the ROV and the cable for the van veen to ensure locational accuracy.

7.3.1. Multibeam Sonar

Swath or multibeam echo sounders were designed to determine the depth of water and the topography of the seabed. Multibeam sonar units use a broad fan-shaped sonar pulse from a transducer to create a swath across a track (perpendicular to the vessel) that records a small along track (the forward movement). Like most geophysical units, the computer software then calculates a two-way travel time for the acoustic pulses utilising a bottom detection algorithm. The speed of sound in water must be calculated for each survey to ensure accuracy. This is conducted using a velocity probe (Kongsberg 2006).

The multibeam sonar used was the Kongsberg EM 3002 multibeam echo sounder. The survey was recorded in both Hypack 2011 (hydrologic processing software) and seafloor information system (SIS) from Kongsberg Maritime. The multibeam was attached to the hull of the *R/V Antonie*. The beam width used was 60 degrees per side for a total of 120 degrees of the seafloor. The area of the seafloor recorded and the resolution of the data changed with the depth. A single 300 kHz sonar head with 254 beams was utilised. A velocity probe from Kongsberg was operated that connected directly with the Hypack software to correct for the speed of sound through the water. The velocity probe was deployed each day and whenever a new survey location was started. DGPS was utilised to establish locational accuracy. The EM 3002 uses beamformed signal amplitudes, which mean the beams are combined in such a way that signals at some angles experience constructive interference while others experience destructive interference. The vertical or depth resolution is one-cm and seafloor resolution can be as precise as five cm, which was unfortunately not utilised during this survey due to the large area covered.

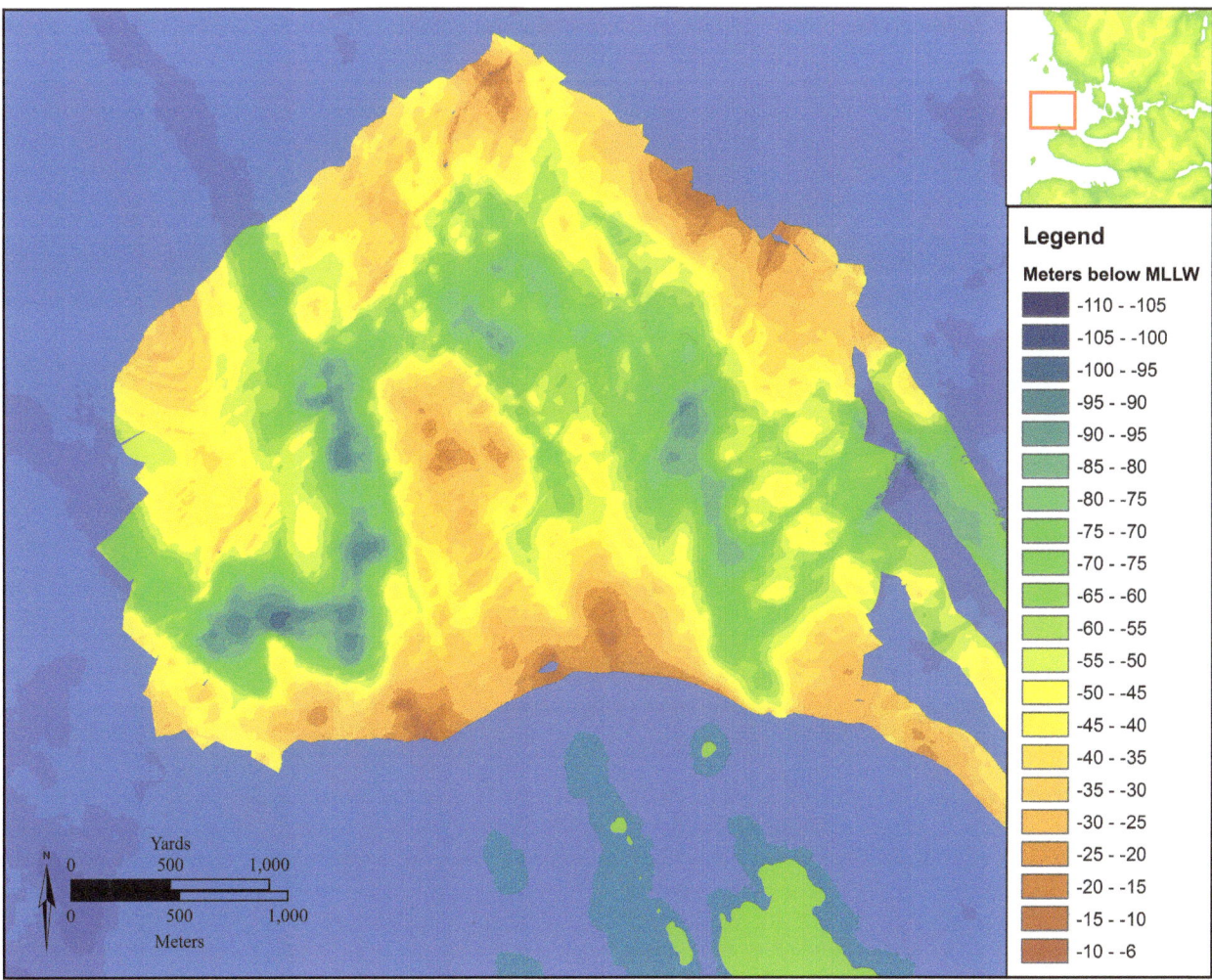

Figure 7-8. Multibeam results at 1.5-m resolution (with five-meter colour palate). The multibeam data extends to the east, where the *R/V Antonie* harboured each night. Data were collected as we approached the survey area.

Uncovering Submerged Landscapes

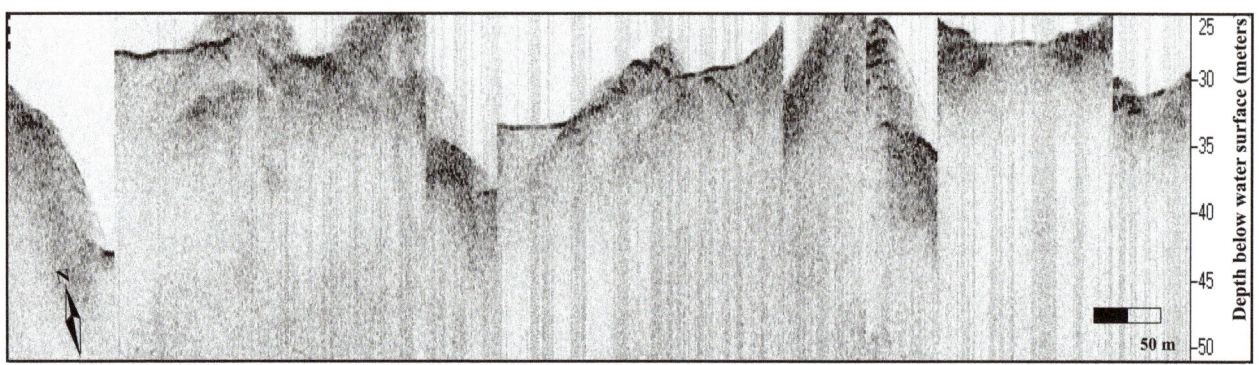

Figure 7-9. Example of the poor quality of sub-bottom data due to calibration issues.

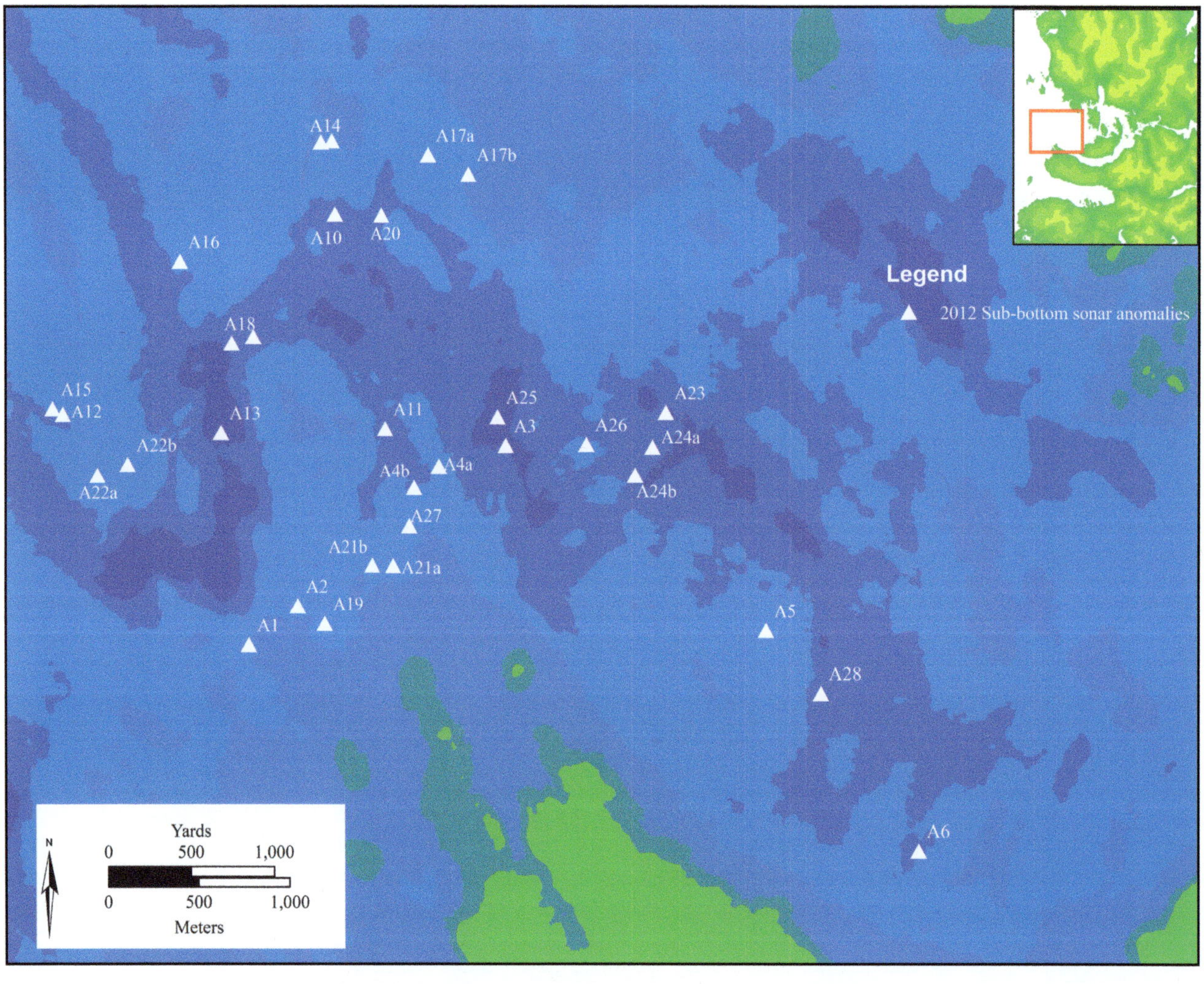

Figure 7-10. Map of identified sub-bottom anomalies. When 'a' and 'b' are included, the anomaly stretches over a large area and the two locations mark the beginning and end of the anomaly or feature, such as figure 7-14.

The EM 3002 includes pitch, roll, and heave stabilisation and records yaw. Pitch is the surge or longitudinal or front/back motion of the vessel. Roll is the sway or lateral or side-to-side motion of the vessel. Heave is the vertical or up-down submotion of the ship. Yaw is the rotation of the vessel. A series of test lines were run to determine average values for pitch, roll, and heave. These test lines were over a flat location and a steep location to record how the ship responded. These values then were inputted into the Hypack software for immediate correction. The SIS dataset was not corrected. The Hypack data were used for analysis. Further post-processing to correct the heave, pitch, yaw, and roll was conducted by Rick Hollar. 'Originally, the software removed a long-term average of the heave from the soundings but I [Rick Hollar] found that not removing that average gave better results' (Hollar 2012 personal communications). Tides in Shakan Bay vary by several meters. Tides were corrected using the predicted tides from the published tide tables for the Shakan Bay Entrance location (Station ID 9450982).

The Hypack data were exported as xyz data by Rick Hollar and inputted into QPS Fledermaus. Using the DMagic tool, the data were combined into a 'pure file magic' or pfm file. The pfm format allows the user to edit points to identify errors and problems with overlapping data. The data were exported various scales from 0.25 m to 1.5 m. Due to the variability in point spacing, the 1.5 m option was chosen for better detail. A combined uncertainty bathymetry estimator (CUBE) was utilised during the generation of the pfm that created uncertainty estimates for the beam points and removes files that are outside normal parameters. Normal parameters are set at five per cent variation in depth. Minimal other post-processing was conducted.

The data were very coarse (figure 7-7). Points are much further apart than was intended for the survey. In the subsequent surveys, the beam angle was narrowed based on depth to maintain a constant resolution of less than 0.5 meter. Currently, some of the data points are greater than one m apart. The 1.5 m resolution is only possible because a 100 per cent overlap was achieved during most of the survey. Some small blue spots in red areas or data holes are visible on the 1.5 m (figure 7-8) multibeam results. This exists when there is less than one multibeam return for a 1.5 m area. No new anomalies have been identified during post-processing of the data. The anomalies identified in the side scan have not been identified in the multibeam data.

7.3.2. Sub-bottom Profiler

Sub-bottom profilers use the same premise as multibeam sonar, but it is a lower frequency and only a single beam. The lower frequency acts like ground penetrating radar (GPR) and provides a two-way sound profile of the sub-seafloor. The profile depicts the density differences between strata and it can identify human-made cultural objects or disturbances such as buried structures or stratigraphic anomalies. The quality of the data is very poor with the sea floor moving throughout the recordings (figure 7-9). Some of the data issues are due to towing the unit behind the *R/V Antonie*, which was resolved when it was mounted on the hull in later surveys.

The sub-bottom profiler used in 2012 was an Edgetech 3200 – SB-216S. The frequency range on this unit is two to 16 kHz. The vertical resolution is between six and 10 cm and it can penetrate from six to 80 m into the sea floor sediments depending on the density of the material. The sub-bottom data were recorded using Edgetech software. This was routed through the Hypack multiple beam recording system for GPS location. The Hypack

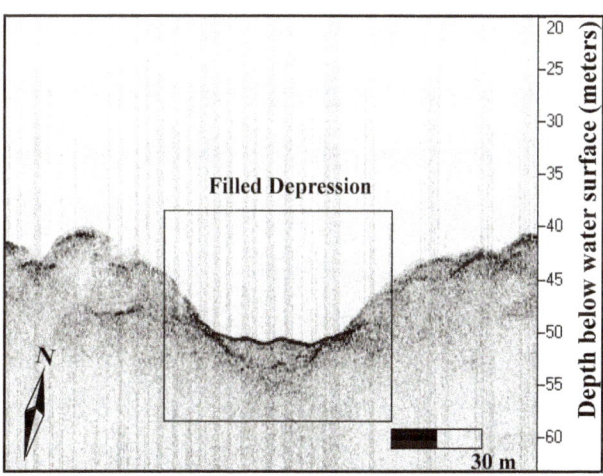

Figure 7-12. Sub-bottom image of a possible filled depression from location A15.

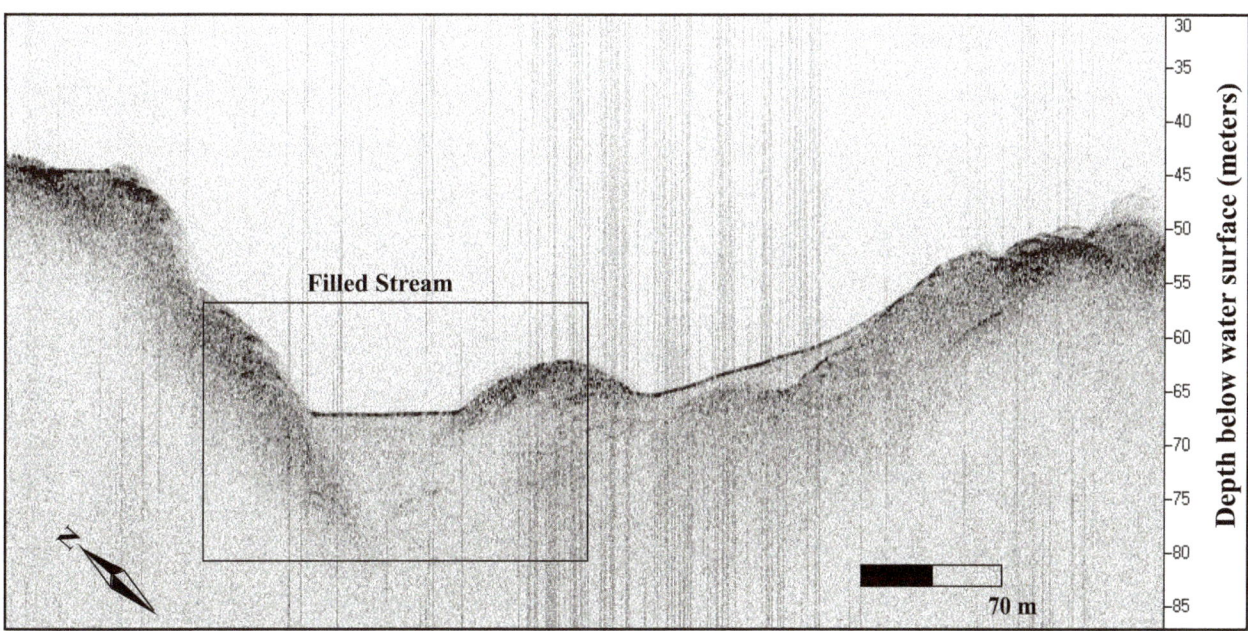

Figure 7-11. Sub-bottom image of a possible filled stream from location A26.

software would often lose the connection to the GPS and would need to be restarted. Each time Hypack was restarted, the settings on the Edgetech software returned to their factory settings (including time). The Edgetech software, Discovery Sub-Bottom, recorded everything in '.jsf' format, which is unique to Edgetech. This software had the ability to convert to '.xtf', which is a universal format, but QPS does not utilise either format of sub-bottom data. QPS required SEG-Y files and the extension for FM Midwater to analyse sub-bottom sonar. The sub-bottom data was analysed in the Edgetech software and has not been referenced to the multibeam or previous side scan data.

Subsequently, the sub-bottom sonar data were analysed independently in Egdetech's Discover Sub-bottom 4.09. A total of 35 possible anomalies were recorded from the sub-bottom data (figure 7-10). Several of these are natural features such as filled streams (figures 7-11 of A26 and figure 7-12 of A15). Much of the data is unusable such as figure 7-9. The scale along the images varies with the speed of the vessel. The indicated scale bar is extrapolated from the GPS data in the sub-bottom profiles. The depth on the right of these images is the depth of the profile adjacent to the scale; any jump in the visual surface results in a change in this scale bar.

Figures 7-13 (A5) depicts a buried sediment change including a higher density layer. There are several examples similar to this including A3, A6, A8, A10, A11, A13, A23, and A25. These could be buried soil horizons with the potential for preserved cultural materials. Figure 7-14 (A9) is a think dense surface return indicating a potential shell deposit. This feature is 130 meters long in the direction the ship travelled. Unfortunately, there is no way without testing to determine if this feature includes cultural deposits.

7.3.3. ROV and Sub-surface Testing

The remotely operated vehicle (ROV) that was used in 2012 was similar to the one used in 2010, but it only had one camera. It was the Sea Eye Marine Falcon. This unit is rated to 300 m depth. It has a high-resolution colour camera with 180 degrees tilt and three variable intensity LED lights totalling 6400 Lumens. The arm has a three-finger gripping function.

In Shakan Bay, a total of 6 hours, 16 minutes, and 57 seconds were recorded of ROV imaging. This does not include time diving the unit or submerging for most dives. Essentially, the time recorded equates to bottom time. The visible field was minimal with the ROV, often only about 46 cm (18 inches). Hypack 2011 and acoustic sensors were used to track the location of the ROV while submerged. A sensor was placed in the water slightly stern of mid-ship on the starboard side and on the ROV unit. This connected to the DGPS that was on top of the F/V Crain to record ROV location.

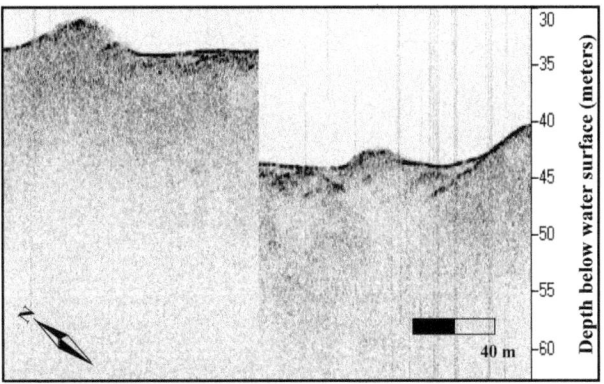

Figure 7-13. Sub-bottom image includes a discontinuous recorded surface with buried sediment layer from location A5.

The van veen grab sampler is a lightweight sampler designed to sample soft sediment. There are two sizes, small and large. Only the small size was available for the 2012 survey (figure 7-15). It weighs 40 lbs. and could penetrate the sediment up to 10 inches. Most of the time, it takes much smaller samples because of its inability to penetrate

Figure 7-14. Sub-bottom image of possible shell deposit from location A9.

Figure 7-15. Van veen grab sampler used for sub-surface testing (Leatherman used as scale).

more consolidated sediment and cannot penetrate the full 10 inches. Additionally, the jaws of the device do not close if a cobble is caught between them resulting in 'blanks' or no sediments to screen. The small van veen was deployed from the *F/V Crain* via a hydraulic winch. It was dropped as quickly as possible and then brought up slowly. The sediment was then dumped into a sled-like plastic bin. The bin was slowly poured onto metal screens. The bin was then rinsed with seawater over the screen to capture any adhering sediment. The sediment was then wet screened using seawater. Two graduated screen sizes were used for each sample: one-half inch and one-quarter inch. Items of interest such as wood, conifer cones, and odd shaped rocks were put in plastic bags for later analysis. Seawater was added to all organic remains to maintain the salinity and moisture levels.

Five areas of Shakan Bay were investigated (figure 7-16): anomalies from the 2010 survey numbered one, two, and three, and two high potential areas referred to as six and seven. The ROV was used to explore the recorded locations for the side scan sonar anomalies at 2010 anomalies one, two, and three. However, there was an unknown and undetermined amount of lag or difference between the recorded GPS locations and where the sonar was when it recorded the location, due to towing the fish behind the vessel. No sub-surface testing was done at area one. One test was conducted at area two that was negative for cultural material.

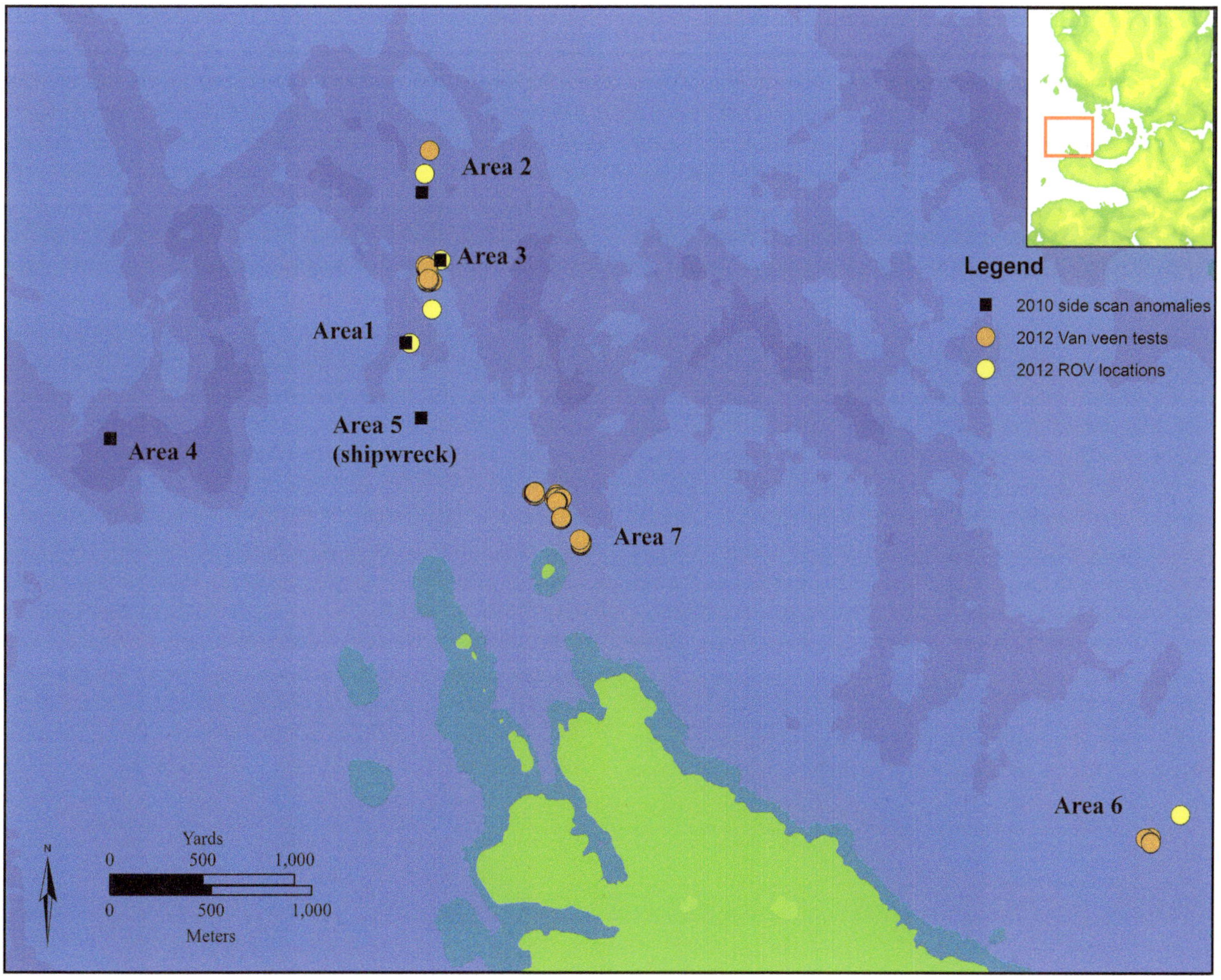

Figure 7-16. Map of ROV investigations or dives and sub-surface test locations.

Figure 7-17. Cut wood from VV25-20 from Shakan Bay Area three.

Figure 7-18. Piece of rounded wood recovered from Area seven (VV-26-06).

A total of 35 sub-surface tests were conducted at area three; of these, 15 were empty. A rock would be caught in the opening of the van veen and all of the sediment would fall out while the device was being raised. One sample, VV25-20 (van veen, 25th of May, sample number 20) included a small piece of cut wood (figure 7-17). A sample of the cut wood was submitted for radiocarbon dating; the date was determined to be modern (CURL-15805). It is suspected that all or most of the ecofacts recovered are modern. It is likely any possible archaeological material is deeper in the sediments than the van veen will penetrate.

Area six was not surveyed in 2010 but was where *the Crain* was anchored on the night of the 23rd of May and was adjacent to a high potential area based on the iteration of the model that existed in 2010. A small black basalt rock was brought up by the ROV at this location. It looked very luminous in the light of the ROV. It was not modified. Nothing of note was identified during this test dive. Four sub-surface tests were conducted. The matrix was muddy sand mixed with shell and some small rocks. No ecofacts or sediment samples were recovered from this location.

Area seven was identified by examination of the multibeam data collected the previous day that suggested the presence of a series of palaeo-beaches. It was also a high potential location on the model. A total of 50 samples were brought up and only one of these was empty. Three locations were sampled within the area. The matrix was a silty-sand with rocks, shell, and shell-hash. The ecofacts recovered include bark, wood, coniferous cones, shale, and irregularly shaped rocks. Some of the bits of wood were rounded, but none of them showed definitive signs of cultural modification (figure 7-18). The rounding was likely naturally caused by weathering and water erosion.

7.4. Summary

The composition of the seafloor was somewhat surprising. There was a significant amount of modern bark, coniferous cones, branches, and even logs. There were small pieces of wood and other organic materials on top of the sediments. The seafloor had a scattering of rock outcrops. The rock outcrops appear to have been scoured, likely by past glaciations. The silty areas were teaming with small sea life including numerous starfish. These starfish create small burrows or beds that could be incorrectly interpreted as small pits or cultural-depressions.

The study region is vast, with large areas of open-ocean. Many areas of open-ocean will be dangerous and difficult to survey due to environmental conditions and the size of the survey vessels, as was experienced in the 2013 survey when numerous attempts were made to survey in the southwest of the study region in the open ocean near Suemez Island. The large area requires the ability to collect and process larger sediment samples than the van veen sampler can provide. Conducting over 50 van veen samples in a single day is significantly less, in sediment volume, than 50 shovel tests (30 cm by 30 cm tests), which is not enough material. Additional field-testing is essential to determining the outcome of the archaeological site high potential model. New methods have been developed using a suction dredge system attached to the ROV to increase sample size and precision on the sea floor. This will also allow the testing of sediments that are not currently on the surface. Shakan Bay was the location that methods were tested for the Gateway to the Americas I and II projects. Subsequent years led to better sub-bottom and multibeam data and the suction dredge developed allowed for over a meter testing in some locations. The process is still very slow compared to terrestrial sub-surface testing. Unfortunately, the opportunity to return to Shakan Bay did not occur during the remaining years of the Gateway to the Americas II project. Funding applications are under development to return to Shakan Bay to test the sub-bottom anomalies and to refine the sea level reconstruction.

8

Pre-10,000 cal BP Archaeological Sites on the Northern NWC

8.1. Introduction

Currently, the author is aware of eighteen reliably dated archaeological sites on the northern NWC pre-10,000 cal BP (table 8-1, figure 8-1). Some of the sites are on raised beach terraces from when sea level was higher than modern (10,600 to 5800 cal BP). Others, especially the mainland BC sites are where sea level has changed little (figure 3-4). These are the types of sites this high potential model is trying to locate. These sites are in the published literature. Table 8-1 has the oldest component for each of the pre-10,000 cal BP sites organised in geographic order from north to south (figure 8-1). Sites with dates prior to 10,000 cal BP were selected for several reasons. First, Baichtal and Carlson (2010) say that by 9,000 cal BP northwest POWI was an 'interaction corridor'. Secondly,

Table 8-1. Average calibrated 14C dates for sites older than 10,000 cal BP (Lee 2007: 32, 46)

Region	Site	Mean of Calibrated Age Ranges
SE Alaska	Ground Hog Bay 2 (JUN-037)	11,528
	Hidden Falls (SIT-119)1	10,157
	Irish Creek (PET-160)	10,484
	On Your Knees Cave(PET-408) - Human remains1	10,207
	On Your Knees Cave(PET-408) - Bone tool	12,129
	Trout Creek Upper Terrace (PET-650)	10,364
	Rice Creek ((CRG-592)	10,235
Haida Gwaii	K1 Cave1	12,650
	Richardson Island1	10,442
	Werner Bay	12,481
	Arrow Creek 1	10,625
	Arrow Creek 21	10,584
	Gaadu Din Cave 11	12,683
	Gaadu Din Cave 21	12,480
	Kilgii Gwaii	10,675
BC Mainland	Far West Point (GcTr-6)	11,045
	Triquet Island (EkTb-9)	13,850
	Kildidit Narrows Site (ElTa-18)	11,419
	Kildidit Narrows Site (ElTa-18) - charcoal but no unequivocal artefacts	13,564
	Namu (ElSx-1)	11,049
	Pruth Bay (EjTa-15)	10,608
	North Kwakshua Channel (EjTa-4)	13,300
1 Average of several mean calibrated age ranges		

10,000 cal BP is 600 years after sea level rose above modern sea level in southeast Alaska. Hence, sites from this period should be visible. Fedje and Mackie (2005, 158) identify the Kingii Complex starts at 8900 cal BP and Dixon (2013a) has the start of the Northwest Coast Microblade tradition in the north of the region by 10,600 cal BP. Finally, as the models were conducted from 11,000 to 16,000 cal BP, this later time-period is the focus of this research. As more sites are identified older than 11,000 cal BP, the range will be further narrowed. Oddly, each of the three regions included in this summary (Southeast Alaska, Haida Gwaii, and BC mainland) each currently have six sites identified.

Each of the sites discussed below is analysed using a 50-meter resolution version of the study's model using modern sea levels. If a more accurate model exists, that will be discussed. Only one site, Trout Creek, is within the two-meter resolution model. The three southern sites in the southeast Alaska are within the 10-meter resolution model. When in doubt, the values are reconstruction based on manual location analysis. See chapter five and specifically table 5-4 for the model weights.

8.2. Southeast Alaskan Sites

Southeast Alaska is a political boundary defined between the United States and Canada by the Alaskan border. This boundary divides the Haida people. It is not reflected by the languages and cultures of the northern NWC.

8.2.1. Ground Hog Bay 2 (11,528 cal BP)

The site is west of Juneau in Glacier Bay National Park. It was located in the summer of 1965 (Ackerman 1968, 55). Ground Hog Bay 2 (JUN-37) is located on a glaciomarine terrace 17-18 m above mean sea level (amsl) on Chilkat Peninsula. Ground Hog Bay is protected from Icy Strait by an offshore reef and headlands to the northwest and southeast (Ackerman 1996, 424). The site is set back from the current beach, 34.5 m inland from high tide (Ackerman 1968, 55). Following local deglaciation (11,500 cal BP) (Ackerman 1968, 56), the terrace formed the edge of an ancient beach, the ocean deposited sand and gravels overlay the till surface. Glaciomarine sediments from elevated terraces on either side of Icy Strait were dated to 16,366 cal BP (13,420±130 (SI-2082)) and 13,511 cal BP (11,630±145 (SI-2113)) (Ackerman, Hamilton, and Stuckenrath 1979, 198). These provide maximum limiting dates for the Ground Hob Bay 2 terrace.

Ackerman (1968) identified component I as the youngest. Component II has eight dates ranging from 4,645 cal

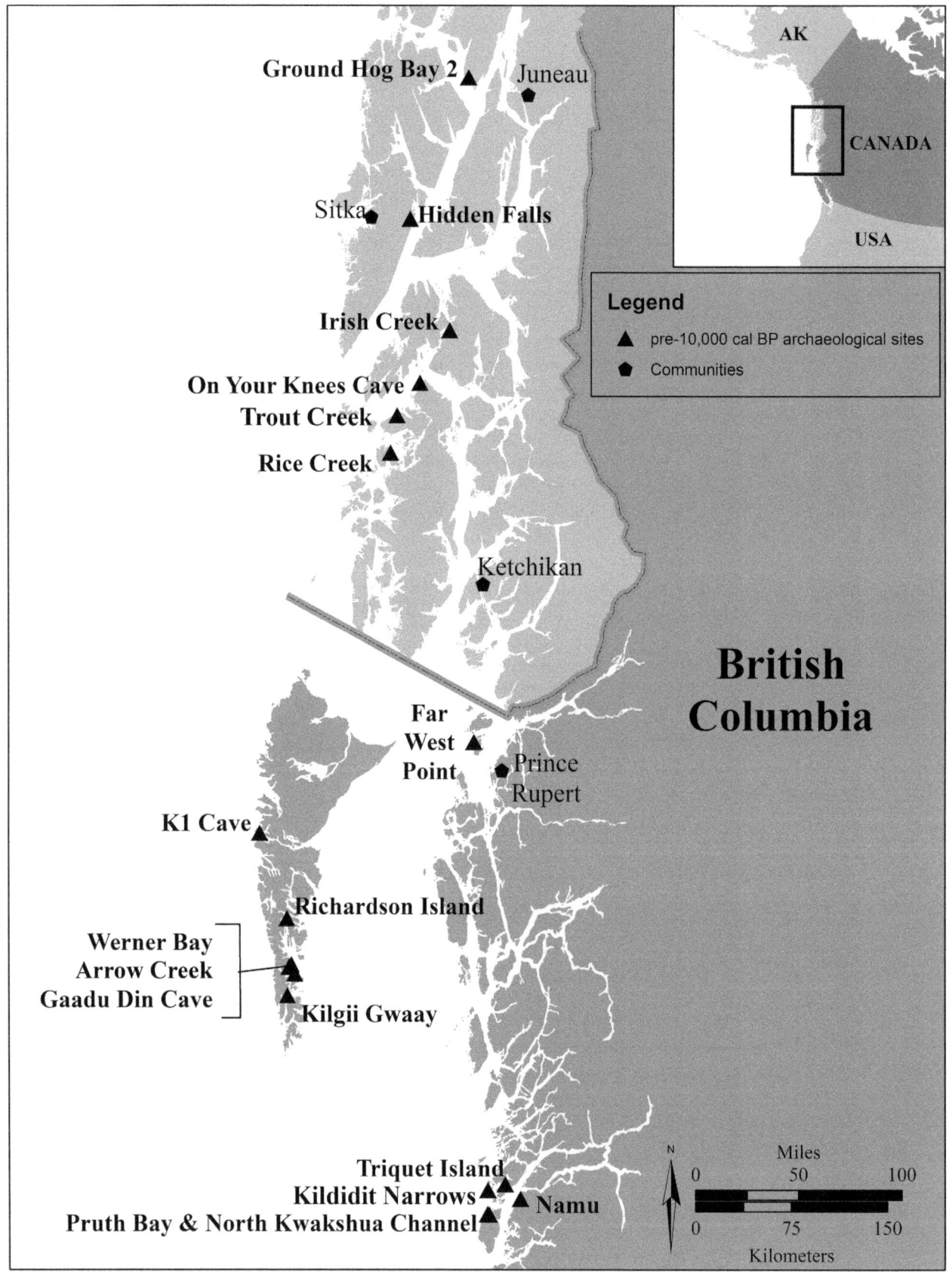

Figure 8-1. Map of Pre-10,000 cal BP archaeological sites from the northern NWC.

BP (4155±95 (I-7056)) to 9916 cal BP (8880±125 (I-7057)) (Ackerman 1996, 426). Component III has three dates 10,259 cal BP (9130±130 (I-6304)), 10,406 cal BP (9220±80 (SI-2122)), and 11,528 cal BP (10,180±800 (WSU-412)) (Ackerman 1996, 426). Due to the large error associated with the last date, it is not regarded as a reliable basal date (Ackerman, Hamilton, and Stuckenrath 1979; Ackerman 1996). Components II and III were later seen as stratigraphically indistinguishable and combined as the lower component of the site (Ackerman 1996, 425–26).

The diagnostic artefact in the assemblage is a wedge-shaped, frontally fluted microblade core. Obsidian is thought to have been imported from Suemez Island or Mount Edziza in British Columbia (Ackerman 1968; Lee 2007). Bifacial fragments were also present. Two obsidian biface fragments came from the lowest part of the site. The lower component (II and III) also includes side scrapers. No hearth or evidence of structure was noted during excavation of the lower component, though there might have been some fire-altered rock (Ackerman 1996, 428–29).

Ground Hog Bay 2 is outside the five-meter and 10-meter resolution model area and was assessed based on ArcGIS's provided topographic maps, Alaska stream data, and data created for the 50-meter resolution NWC model. The site has a slope of zero degrees and an aspect of 161 degrees or south-southeast. The slope would be rank five and the aspect would be rank two. The coast is 34.5 m from the site, which is a rank of five. The nearest coastal sinuosity is one which is medium potential for a site. The site is approximately 300 m from a stream, which is a rank of five. There are no lakes within 3000 m, so the rank would be zero. The junction of a stream and the modern coast is 300 m away, which is a rank of four. The nearest archaeological site is over seven km away; the rank would be zero. Putting this math together using the same system as the weighted overlay function and weights as the final model, the location where Ground Hog Bay 2 is located would have a weight of three. This means it would be a moderately high archaeological potential.

8.2.2. Hidden Falls (10,157 cal BP)

The Hidden Falls (SIT-119) site is approximately eight m amsl on a point at the head of Kasnyku Bay on northeast Baranof Island, northeast of Sitka. It was excavated in 1978 and 1979 (Davis 1989, 23, 27) and was discovered as a result of road construction for a fish hatchery.

Hidden Falls is a multicomponent, stratified site at which Davis identified three components. Component I is the oldest and the only one relevant to this research. Component I was found in stratigraphic unit (SU)-I, which is a discontinuous palaeosol, and is found in units SU-G, -H1, -H1a, and -H2, which are Holocene ablation tills (Davis 1996, 416–17). A total of 14 radiocarbon dates were reported for Component I. Eleven dates are from SU I, and ten of these were on wood. The eleventh was on wood and is younger than the other dates 8,000 cal BP (7175±155, SI-3777). Three radiocarbon samples were from H2. Two of these were wood and one was charcoal. The wood dates range from 12,240 cal BP (10,345±95 (SI-4360)) to 8,774 cal BP (7,900±90 ^{14}C years (SI-4340)) and include three dates over 11,500 cal BP (10,000 ^{14}C years). Only four of the dates are stratigraphically younger 10,000 cal BP, including the diluted wood. The charcoal from H2 was reported a date of 10,157 cal BP (9060±230 (Beta-7440)). Davis combines the radiocarbon dates and dates from pollen cores from Muskeg Lake to determine that the occupation was around 10,800 cal BP (9,500 ^{14}C years). The date of 10,157 cal BP is calculated from the archaeologically significant published radiocarbon dates.

The artefact assemblage consisted of a unifacial tool, microblades, split-cobble and split-pebble tools, choppers, hammer stones, a variety of scraping tools, gravers, an abrader, burins, and utilised flakes. A total of 612 lithic artefacts and debitage were recovered from component I in the two years of excavation. A biface fragment was recovered, but Davis (1996, 417) questions its provenance because it was associated with a tree stump removal. Obsidian was from both Mount Edziza in British Columbia and Suemez Island, Alaska.

Hidden Falls is outside the five-meter and 10-meter resolution model area and was assessed based on ArcGIS's provided topographic maps, Alaska stream data, and data created for the 50-meter resolution model. The slope is four degrees and the aspect is 129 degrees or southeast. This means the slope rank is four and the aspect is a rank of one. The coast is less than 70 m away, which is a rank of five. The sinuosity is two, which is medium potential for an archaeological site. The nearest stream is less than 100 m away, which is a rank of four. The nearest lake is approximately 590 m away, which is a rank of three. The nearest tributary junction is 270 m away, which is a rank of four. There is a site located 2.7 km away; therefore, the distance from site rank is one. The overall weight would be three, or moderately high archaeological potential.

8.2.3. Irish Creek (10,484 cal BP)

Irish Creek (PET-160) is located on an elevated terrace 1.3 km inland from the mouth of Irish Creek on the west side of Kupreanof Island. Original excavations at the site by Roberts in 1981 and 1984 dated the site to 2,317 cal BP (2240±70 ^{14}C years (Beta-02829)). Moss and colleagues used obsidian hydration to date the site to 5000±500 years before present (Moss et al. 1996). Carlson and Baichtal (2015) returned to the site as part of the raised terrace study and were able to collect four charcoal samples from the lowest cultural levels. These samples date between 10,196 cal BP (9280±40 C^{14} years BP (Beta-282535)) and 10,543 cal BP (8990±40 C^{14} years BP (Beta-282533)) (R. J. Carlson and Baichtal 2015, 131). Irish Creek's artefacts include microblades that correspond with Dixon's (2013a) chronology for the region. Other artefacts include cores, retouch flakes, scrapers, gravers, drills incised pebbles, and unmodified pebbles (Moss et al. 1996, 79).

Irish Creek also outside the modelled areas and was assessed the same as Ground Hog Bay 2 and Hidden Falls. The slope is zero degrees, which is a rank of five. The aspect is 243 degrees or west-southwest, which is a rank of two. The site is less than 100 m from a stream, which is a rank of four. There are no lakes within three km, which is a rank of zero. The coast is 700 m away, which is a rank of three. The sinuosity is 1.4, which is medium potential for an archaeological site. The nearest tributary junction is 700 m away, which is a rank of three. There are no sites within five km, which is a rank of zero. The overall weight for Irish Creek would be three, or moderately high archaeological potential.

8.2.4. On Your Knees Cave (10,207 cal BP and 12,129 cal BP)

Cavers located On Your Knees Cave (OYKC - PET-408) in 1993. The cave is located in the northwest area of POWI, near Point Protection. The cave is approximately 135 m amsl. It was not in a coastal setting (Dixon et al. 1997, 707). The cave has a total length of 70 m and consists of two passages: Bear and Seal Passages. OYKC was used as a carnivore den throughout the late Pleistocene and the Holocene (Heaton and Grady 2003). The site was subject to a variety of depositional events including the accumulation of unsorted inorganic sediments that could have been deposited by glacial melting and landslide events. The cave contained brown bear *(Urus Arctos)* and black bear *(Urus Americans)* remains dating to before the LGM and ringed seal (*Phoca hispida*) remains dating to the LGM. The vertebrate remains indicate that the region was an ice-free refugium and capable of supporting humans and animals during the LGM. OYKC is useful for defining vertebrate faunal history for POWI over the past 50,000 years. The palaeontological work focused on the inside of the cave while the archaeology primarily focused around the cave entrance (Heaton and Grady 2003, 22–23).

In 1996, Timothy Heaton found part of a human pelvis in OYKC. Excavations stopped immediately when human remains were recognised, complying with North American Grave Protection and Repatriation Act (NAGPRA). With tribal approval, the excavation continued. The OYKC human remains represent a man who died in his mid-twenties, based on the partial eruption of his wisdom teeth. The remains are comprised of a mandible in two pieces (including teeth except for incisors), a right pelvis, cervical vertebrae, lumbar vertebrae, three thoracic vertebrae, three ribs, three incisors, and one canine (Kemp et al. 2007, 606; Lindo et al. 2017). The mandible dated to 11,283 cal BP (9730±60 (CAMS-29873)) and the pelvis dated to 10,260 cal BP (9880±50 (CAMS-32038)) (Dixon et al. 1997, 703). The human remains, named Shuká Káa by local Haida people, had a ^{13}C value of -12.5 per cent. This suggests a diet principally based on marine resources (Dixon et al. 1997, 703; Lindo et al. 2017). The calibrated values incorporate the marine correction for each bone using the marine reservoir correction of ΔR=495 ± 24 and Marine09 in Calib6.0.

The human remains are contemporary with the cultural occupation of the site. Three dates were run on charcoal: 9739 cal BP (8760±50 (CAMS-43991)), 10,377 cal BP (9210±50 (CAMS-43990)), and 10,330 cal BP (9150±50 (CAMS-43989)). Also recovered from this cultural layer were obsidian microblades, bifaces, and other tools. These tools are consistent with the NWC Microblade tradition (Dixon 1999, 118–19, 2013a). Some researchers have started to refer to OYKC as Shuká Káa Cave (R. J. Carlson and Baichtal 2015), but it is still technically named OYKC.

The only artefact recovered near the human remains was a piece of mammal bone that was modified, possibly as a flint flaker. It exhibited similar staining and preservation as the human remains and is possibly associated with the human remains. This piece of a mammal bone dates to 12,129 cal BP (10300 ± 50 (CAMS-42381)) (Dixon et al. 1997, 705–6).

OYKC is located within the 10-meter resolution model, so the variables used in the model can be used to describe this site. As the model was only generated from 10,500 cal BP and older, the 10,500 and 12,000 cal BP models will each be discussed for this site. Unfortunately, OYKC is a limestone cave, which is a feature that this model does not intend to locate.

Using the 10,500 cal BP model data, the final weight is two. The slope is ranked four. Aspect is ranked two. Streams are ranked zero. Lakes are ranked zero. The sinuosity is low and coastal distance is ranked four. Tributaries are ranked zero. Sites are ranked five for distance to site.

Using the 12,000 cal BP model, OYKC is also ranked two. Slope is ranked two. Aspect is ranked one. Streams are ranked two. Lakes are ranked two. Tributaries are ranked two. Sinuosity is low and ranked three. Again, distance from sites is ranked five. These higher rankings for streams and tributaries are due to the lower sea level, more streams and tributaries that would have formed. Finally, the final model for the region, which encompasses 10,500 to 16,000 cal BP, has OYKC weighted as three or moderately high archaeological potential. So, despite the model not focusing on caves, the site is still identified by some of the models.

8.2.5. Trout Creek Upper Terrace (10,364 cal BP)

Trout Creek Upper Terrace (PET-650) is the older raised terrace feature along Trout Creek. The lower terrace (PET-651) is 11 meters above mean low low-water and dates to 3970 cal BP. The upper terrace, which is the relevant site here located on an 18 meter above mean low-low water terrace and dates from 9682 cal BP (8730±50 (Beta-264082)) to 10364 cal BP (9130±40 (Beta-288260)) (R. J. Carlson and Baichtal 2015, 128). The site is located on the west side of Kosciusko Island just south of Shakan Bay. Microblades, bifaces, and flaked tools were all recovered from this site. The dates suggest a single continuous occupation at this location.

Trout Creek Upper Terrace is located on both the two-meter and 10-meter resolution models. The model results are three for all four models analysed in chapter six (study region deep, study region modern, study area 11,000 cal

BP, and study area deep). Using the same calculations as used for other sites outside the model, Trout Creek Upper terrace has a slope of five degrees, which is ranked at three. The aspect is 130 degrees or southeast, which is ranked two. The site is 468 m from the coast, which is ranked five. The sinuosity of the nearby coast is 1.3, which is medium for archaeological site potential. The nearest stream is 75 m away, which is ranked four. There are no lakes within three km, which is ranked zero. The nearest tributary junction is 468 m away, which is ranked four. The nearest site is 185 meters away, which is ranked three. The overall weight also adds to three, which is moderately high archaeological potential.

8.2.6. Rice Creek (10,235 cal BP)

Rice Creek (CRG-592) is part of a series of sites located along Rice Creek on the north side of Heceta Island. Artefacts recovered include microblades made of obsidian, chalcedony, and fine-grained cryptocrystalline silicate. The site dates between 7143 cal BP (6100±50 (Beta-264579)) and 10,235 cal BP (9090±50 (Beta-264554)) (R. J. Carlson and Baichtal 2015, 128–29).

Rice Creek is within the 10-meter resolution model where the model results are three for both the study region deep and study region modern. Using the values calculated for the sites outside of the study area and region, Rice Creek has a slope of four degrees, which is ranked as four. The aspect is 55.7 degrees or between northeast and east-northeast, which is a rank of two. The site is 45 m from a nearby stream, which is a rank of four. There are no lakes within three km, which is a rank of zero. The coast is 220 m away, which is a rank of four. The sinuosity of the nearby coast is 1.3, which is medium archaeological potential. The nearest tributary junction is 76 m away, which is a rank of five. The nearest site is 207 m away, which is a rank of three. The overall weight adds up to three or moderately high archaeological potential, as was derived from the models.

8.3. Haida Gwaii

Haida Gwaii (Queen Charlotte Islands) is outside the five-meter and 10-meter resolution models. The NWC 50-meter resolution model is of limited value due to the unknown precise locations of many of the Haida Gwaii sites. Some of these sites are not in the data provided from the BC archaeology branch in 2009, nor has the author been able to reproduce precise locations for all the sites based on published maps. Similar data sets were used to evaluate these sites as were used for Ground Hog Bay 2, Hidden Falls, and Irish Creek, when possible.

8.3.1. K1 Cave (12,650 cal BP)

Located at the head of a fjord on the west coast of Moresby Island, K1 cave is a solution cave in limestone (Ramsey et al. 2004, 104). Within the cave, black bear (*Urus Americans*) remains were recovered dating from 10,612 cal BP (9,376±50 (CAMS-79488)) to 17,507 cal BP (14,390±70 (CAMS-75746)). CAMS-75746 was corrected for marine reservoir effect by -150 years based on the isotopic signature. The ^{13}C isotopic values for the bears support a terrestrial diet except for the oldest, which was a mixed marine-terrestrial signature (Ramsey et al. 2004, 107–8). The bases of two large, foliate points were recovered in association to these black bear bones and may derive from bear hunting (Fedje et al. 2008). One point was recovered from a level overlain by sediments dated to 12,488 cal BP (10,510 ^{14}C years) and underlain by a date of 12,812 cal BP (10,960 ^{14}C years). The other point was bracketed by dates of 12,500 to 12622 cal BP (10,525 and 10,660 ^{14}C years). Based on the stem width and heavy lateral and basal grinding, it is likely that these points were end-socket hafted. Both projectile points were made from a creamy-white to yellowish-brown chert (Fedje et al. 2008, 20). This site was not listed in the provided BC archaeology branch data. An exact location is not known for this research and an estimate from published maps is not accurate enough to test the model. This predictive model is not trying to reconstruct cave site locations.

8.3.2. Richardson Island (10,442 cal BP)

The site was located at a secondary deposit in the inter-tidal zone on the west side of Richardson Island in Darwin Sound. The primary deposit was located on a raised marine terrace 15-16 m amsl. Over four vertical meters of stratified deposits accumulated over 1000 years at this site, and more than fifty deposition events were identified, at least twenty of this contained evidence of human occupation (Fedje 2003, 31; Fedje et al. 2018, 41). Fourteen radiocarbon dates were run on charcoal from the raised beach. The underlying diamicion dated to 10,943 cal BP (9590 ±50 ^{14}C years (CAMS-39877)). The site dates from 10,442 cal BP (9290±50 (CAMS-39875 & CANS-39876)) to 9477 cal BP (8490 ±70 (CAMS-16199)) (Fedje 2003, 33). A few grams of calcined bone, mainly of fish, were recovered. The lithic assemblage consists of over 8300 artefacts from the 1995 and 1997 excavation and over 5000 artefacts from the 2001 excavation. There are two components at the site: [1] a Kingii complex (10,500 to 10,250 cal BP (9300 to 9000 ^{14}C years)) and [2] a subsequent Early Moresby tradition occupation dating from 10,250 to 9600 cal BP (9000 to 8500 ^{14}C years) (Fedje 2003, 33).

Large foliate points with slight basal edge grinding and contracting stems, referred to as Xil points, were recovered including two complete points, eleven base fragments, several point tips, and a number of mid-sections. Thirteen Xil points date from 10,500 to 10,000 cal BP (9300 to 8900 ^{14}C years) (Fedje et al. 2008, 21). These are mainly in component one of the site, which lacks microblade technology. There was only one chopper or core tool found in this component (Fedje and Christensen 1999, 645).

Component two is characterised by Xilju points. These are small contracting stem points that date from 9900 to 9800 cal BP (8800 to 8700 ^{14}C years) (Fedje et al. 2008, 22).

Microblades are abundant in all post-10,000 cal BP (8900 ^{14}C years) strata, both tabular and conical types. Bifaces decrease in abundance with time. The assemblage includes core scrapes, cobble choppers, and both unidirectional and multidirectional cores (Fedje and Christensen 1999, 645). The presence-absence of microblades at this site corresponds to Dixon's (2013a) north to south transition from Pebble Tool tradition to Northwest Coast Microblade tradition.

The location was determined based on published maps and the variables were derived from various data sources similar to Ground Hog Bay 2, Hidden Falls, and Irish Creek. The slope was calculated at zero degrees, which is a rank of five. The aspect was determined to be 249 degrees or west-southwest which is a rank of two. The site is 60 m from a stream, which is a rank of four. Richardson Island site is 1528 m from a lake, which is a rank of two. The coast is 30 m away, which is a rank of five. The sinuosity of the nearby coast is one, which is medium potential for an archaeological site. The nearest tributary junction is around 360 m away, which is a rank of four. The nearest site is almost four kilometres away, which is a rank of one. The overall weight for the site is three or moderately high archaeological potential.

8.3.3. Werner Bay (12,481 cal BP)

This is the only underwater prehistoric site currently known on the northern NWC. It consists of a single possibly utilised flake that was recovered during a marine survey in Werner Bay, next to Matheson Inlet and Juan Perez Sound. The flake was found on a drowned delta floodplain at a depth of 53 m. There is an utilised edge of the flake, which is still sharp. Use-wear analysis suggests the flake was used as a knife for cutting meat or soft vegetables. The site is dated to around 10,000 BP based on sea level history. Radiocarbon dating of the barnacle attached to the flake returned 12,481 cal BP (10500±40 ^{14}C years (CAMS50947)) as a minimum date that the flake was deposited on the seafloor (Fedje and Josenhans 2000, 101–2; Fedje 2000; Josenhans et al. 1997; Fedje, Magne, and Christensen 2005). This site was not listed in the BC archaeology branch database. An exact location is not known for this research and an estimate from published maps is not accurate enough to test the model. The location of this flake is harder to determine than the other sites and as such, it is unclear if this isolated flake would have been identified by the model. Additionally, as the flake was 'grabbed', not excavated in-situ, the location may be in a secondary location.

8.3.4. Arrow Creek (10,625 cal BP and 10,584 cal BP)

There are two relevant sites on Arrow Creek. They are site 1 and 2. Site 1 is located on an alluvial fan along Arrow Creek I in Matheson Inlet, Juan Perez Sound. It is 200 m upstream from the river mouth and 15 to 17 meters a. It was first recorded by Hobler in 1976 (Fedje, Magne, and Christensen 2005, 217–18; Hobler 1978). Ten radiocarbon dates have been run on organics from this site. They range from 10,625 cal BP (9390±60 ^{14}C years) (CAMS-8375) to 6540 cal BP (5650±70, 5650±90, CAMS-4112 and CAMS-4111). Several radiocarbon dates were older prior to marine reservoir correction (Fedje et al. 1996, 134: table 1). Though the site is stratified, the only organic remains are some small amounts of charcoal (Fedje, Mcsporran, and Mason 1996, 140). Thirty-five lithic artefacts were recovered. There were no microblades or microblade cores found, but microlithic cores and blade-like flakes were recovered (Fedje, Mcsporran, and Mason 1996, 122).

Arrow Creek 2 was located by a Haida archaeologist, Captain Gold, as a result of his discovery of stone tools on raised marine sediments. The site is at the head of the estuary near the modern tidal limit. It dates from 11,578 cal BP (9900±90 (CAMS-9968)) to 10,088 cal BP (9010±160 (CAMS-8377)) (Fedje et al. 1996, 134: table 1) and was inundated around 10,250 cal BP (9000 ^{14}C years) by rising sea level (Fedje et al. 1996, 138). There were 110 lithic artefacts and two features identified at the site. Only one diagnostic artefact was found; it is an exhausted conical agate microblade core. Other lithic artefacts include debitage and unifacially modified flake tools. The two features are a hearth and a low wall of cobbles that Fedje and colleagues (1996a: 139) suggest may have been an old fish trap. The wall is a combination of larger rocks with smaller cobbles positioned in spaces. It is approximately 40 cm high (Fedje, Mcsporran, and Mason 1996, 129).

Arrow Creek was in the BC archaeology branch database, unfortunately, it is not the same site as Fedje and Josenhans (2000) have mapped. The location has a slope of one degrees and an aspect of 40 degrees or northeast. Slope would be ranked five. Aspect would be ranked one. The site is just over 200 m from the coast. The distance from coast rank is five. The coast is moderately sinuous. The site is approximately 200 m from a stream and a tributary. This is based on data from the Canadian government and GeoBase. According to maps by Fedje and Josenhans (2000), the site is on the stream. This is an error that is very common is site locations and explains the need to have the highest potential for stream be the 500 m buffer not the 100 m buffer. Streams would be ranked five. Tributaries would be ranked four. The site is 600 m from a lake; lakes would be ranked three. The site is 500 m from two different shell middens. The distance from site rank would be three. This means the Arrow Creek sites location is moderately high archaeological potential with a weighted value of three.

8.3.5. Gaadu Din Caves (12,683 cal BP cave 1, 12,480 cal BP cave 2)

Gaadu Din cave 1 and 2 are on Burnaby Island. They are on the east coast of Moresby Island, and both have produced vertebrate faunal remains. Two large foliate points with contracting stems lacking significant edge grinding dating to approximately 11,500 cal BP were recovered from Gaadu Din Cave 1. These points are typologically classified as being intermediate between the Taan (found at K1 and Gaadu Din Cave 2) and Xil styles (found at Richardson Island) (Fedje et al. 2008, 21).

A complete spear point was recovered from Gaadu Din Cave 2. It was made from a yellowish-brown chert and was heavily ground on the lateral stem margins. The base is broad, rounded, and unground. It is similar to the spear point bases found in K1 cave and is stratigraphically associated with faunal remains dating to 11,922 cal BP (10,220 ^{14}C years) (Fedje et al. 2008, 20–21).

This site was not listed in the BC archaeology branch database. An exact location is not known, and an estimate from published maps is not accurate enough to test the model. As these are cave sites, these locations are not designed to be identified by the model.

8.3.6. Kilgii Gwaay (10,675 cal BP)

This is an intertidal site located in a small embayment on the south side of Ellen Island at the southernmost tip of Haida Gwaii (Fedje 2003, 34; Fedje et al. 2001, 98). The site was identified in 1991 and 1500 lithic artefacts were collected from the surface. In 2000, the site was mapped, and two 50 cm square test units were excavated. Stone tools, faunal remains, and waterlogged wood were recovered. A bone splinter was dated to 10,686 cal BP (9460±50 (CAMS70704)). Further work was conducted in 2001 and 2002 (Fedje 2003, 35).

Kilgii Gwaay was occupied from 10,625 to 10,675 cal BP (9450 to 9400 ^{14}C years), during which time sea level was a few meters below modern sea level. The site was flooded in 10,625 cal BP. The archaeological site extends one to three meters below modern high tide and is distributed in a horseshoe pattern around a former pond or lagoon. Marine transgressions have disturbed most of the site, but there is a small area of in situ archaeological deposits on the west side of the basin (Fedje, Mackie, et al. 2005, 187).

Twenty-three radiocarbon dates have been run for Kilgii Gwaay. Twelve of the dates are from archaeological deposits, including nine from stratigraphically paired samples of shell and wood and two from wooden artefacts and one from a spirally fracture bear bone. Two dates are from the intertidal area, one date is from the underlying palaeosol (Fedje, Mackie, et al. 2005, 195). From a total of 16 m^2 of excavation, there were thirty-nine vertebrate and 16 invertebrate taxa. Over 4,200 vertebrate specimens that included 13 taxa of fish, six of mammals, and 20 of birds were recovered. Fish are dominated by rockfish, dogfish, and lingcod while mammals are predominately bear, harbour seal, and sea otter. Eighty-four per cent of the shellfish weight was California mussel. Approximately 4,000 lithic artefacts were recovered. The formal tools are unifacial including a small unifacial stemmed point. There was a single biface fragment (Fedje, Mackie, et al. 2005, 196). Ninety bones 'exhibiting evidence of human modification' (Fedje, Mackie, et al. 2005, 197), and over 100 wooden artefacts (Fedje, Mackie, et al. 2005, 198) including cordage and wooden tools (e.g. wedges and stakes or pegs) (Lee 2007, 60). The bone artefacts included three awls, a unilaterally barbed point, and a bone billet flaker (Fedje et al. 2001, 106). This site provides an important glimpse into the preservation potential of water-saturated sites and evidence of well-developed and diverse maritime subsistence economy at this early time. Additionally, this site demonstrates that submerged archaeological sites can be found intact within the region even after marine transgressions.

This site was not listed in the BC archaeology branch database and the location was determined based on published maps. The slope was determined to be zero degrees, which is a rank of five. The aspect is 289 degrees or between west and west-northwest, which is a rank of two. There is a stream around 213 m away, which is a rank of five. There is a lake 750 m away, which is a rank of three. The site is on the coast, zero meters away, which is a rank of five. The sinuosity is one or medium archaeological potential. The nearest tributary junction is 220 m away, which is a rank of four. There is an archaeological site 1100 m away, which is a rank of two. The overall weight for Kilgii Gwaay is three or moderately high archaeological potential.

8.4. Mainland British Columbia

The six sites located on the BC mainland are in two geographic areas. The first location, Far West Point, is in the Dundas Island chain near Prince Rupert. The remaining sites are near the town and site of Namu. Many of these sites have only been discovered in the last ten years and many more are anticipated as funding continues to allow further research.

8.4.1. Far West Point (11,045 cal BP)

Far West Point or GcTr-6 is the oldest of the palaeoshoreline survey sites found near Prince Rupert on Dundas Island. The earliest radiocarbon dates are from a shell-free deposit below a shell layer and date between 10,2885 and 11,204 cal BP (UCIAMS-28008) (Letham et al. 2018). Though the authors notes this date may be questionable, as it has not been reproduced. The site contains lithics, faunal, and charcoal. A shell midden was deposited between 7000 and 8000 cal BP (Letham et al. 2018).

The location of the site was reconstructed based on published location data. The slope was determined to be zero degrees, which is a rank of five. The aspect was determined to be 100 degrees or between east and east-southeast, which is a rank of two. The site is approximately 300 m from a stream, which is a rank of five. The nearest lake is 2300 m away, which is a rank of one. The coast is 25 m away, which is a rank of five. The sinuosity of the nearby coast is 1.4, which is medium for archaeological potential. There is a tributary junction approximately 350 m away, which is a rank of four. There is a site 2500 m away, which is a rank of one. Overall, Far West Point would have a weight of three or moderately high archaeological potential.

8.4.2. Triquet Island (13,850 cal BP)

The Triquet Island site (EkTb-9) is currently the oldest dated site on the northern NWC. It is located on Triquet Island west of Namu, just north of Vancouver Island. The earliest component includes lithics, faunal remains, and charcoal. Originally, the site was dated between 11,285 and 11,396 cal BP (9960±25 ^{14}C years (UCIAMS-118001)) (McLaren et al. 2014, 154). A sample recovered from a hearth feature in 2016 was dated between 13,613 and 14,086 cal BP. Unfortunately, the radiocarbon date has not yet been included in a peer review publication (Muckle and Gauvreau 2017). Triquet Island site is a deeply stratified site including preserved wood artefacts dating between 4400 and 7300 cal BP and up to five meters of Holocene cultural deposits (McLaren et al. 2014).

The site location was determined based on published maps and reports. The slope is two degrees, which is a rank of four. The aspect is 24 degrees or north-northeast, which is a rank of two. The nearest stream is almost 700 m away, which is a rank of three. The coast is 45 m away, which is a rank of five. The sinuosity of the nearby coast is 1.3, which is medium for archaeological site potential. The nearest tributary is almost 700 m away, which is a rank of three. There is another site within 300 m, which is a rank of three. The overall weight for the site is three or moderately high archaeological potential.

8.4.3. Kildidit Narrows (11,419 cal BP and 13,564 cal BP)

Kildidit Narrows (ElTa-18) is a shell midden site located on Hunter Island west of Namu and north of Vancouver Island. Cultural remains include stone tools, charcoal, and faunal material. The cultural later has a date of 10,701 to 10,757 cal BP (9490± 20 ^{14}C years (UCIAMS-117997)). There are later Holocene archaeological materials above this date. Below this date, there is a palaeosol with no unequivocal artefacts that date between 13,454 and 13,673 cal BP (11,720±80 ^{14}C years (UCIAMS-118046)). Both dates are listed for this site, as the oldest date is questionable (McLaren et al. 2014, 154).

Kildidit Narrows was in the data provided by the BC archaeology branch. The slope is zero degrees, which is a rank of five. The aspect is 332 degrees or north-northwest, which is a rank of two. The nearest stream is approximately 300 m away, which is a rank of five. The nearest lake is almost 700 m away, which is a rank of three. The coast is approximately 45 m away, which is a rank of five. The sinuosity of the nearby coast is one, which is medium for archaeological site potential. The nearest tributary junction is almost 400 m away, which is a rank of four. There is a site 225 m away, which is a rank of three. The overall weight for this site is four or high archaeological potential. It is the only site analysed here that scores in the high potential category, reinforcing the evaluation of values of three and four having archaeological potential.

8.4.4. Namu (11,049 cal BP)

Namu is the only pre-9000 cal BP site on the BC mainland on the BC mainland of the northern NWC. Namu is considered by most researchers to be part of the central NWC (R. L. Carlson 1979), but Ames and Maschner (1999) note that it is a transitional location with aspects of both the central and northern NWC. It is a seasonally occupied fishing camp at the junction of the Burke Channel and Fitzhugh Sound. Hester and Nelson originally excavated it in 1969 and 1970. They uncovered an occupation sequence spanning 9000 years. Work in 1977 and 1978 extended the sequence back to 9700 BP (R. L. Carlson 1996, 83). Occupations have been divided into six periods, of which only the oldest (period 1) is relevant for this research. Period 1 has been further subdivided into three sub-periods: 1A from 11,480 to 10,160 cal BP (10,000 to 9000 ^{14}C years); 1B from 10,160 to 8900 cal BP (9000 to 8000 ^{14}C years); and, 1C from 8900 to 6800 cal BP (8000 to 6000 ^{14}C years) (Carlson 1996: 91). These are based on four radiocarbon dates older than 10,160 cal BP (9000 ^{14}C years) all from charcoal, 9600 cal BP (8570±90 ^{14}C years (WAT-516)), 10,090 cal BP (9000±140 ^{14}C years (WAT-519)), 10,232 cal BP (9140±200 ^{14}C years (Gak-3244)), and 11,049 cal BP (9720±140 (Wat-452)) (R. L. Carlson 1996, 84: Table 1). The faunal remains are continuous from 6800 cal BP (6000 ^{14}C years) to present but are rare before then. The remains are dominated by fish, specifically salmon (A. Cannon 1998).

The lithic assemblage includes both local and imported stones. Period 1A has 32 lithic artefacts that were recovered from test pits and the river mouth trench which include a unifacial chopper, a foliate biface, three biface knives, and three biface preforms. There are no microblades in period 1A. Period 1B has microblades, ground stone tools, many bifacial tools, and unifacial choppers. There is a total of 206 lithic artefacts from period 1B in the test pits and river mouth trench (R. L. Carlson 1996, 91).

Namu is one of the sites that are listed in the database from the BC archaeology branch. The slope is 0 degrees and the aspect is 279 degrees or west. That means the slope is ranked four and the aspect is ranked one. The site is 79 m from the coast, which is a rank of five. The sinuosity near the site is 1.1, which is medium for archaeological potential. The site is less than 500 m from a stream (Namu Creek), which is ranked five. The nearest lake is 1070 m away, which is a rank of two. The nearest tributary junction is approximately 290 m away, which is a rank of four. The site is approximately 800 m from the nearest known archaeological site, a shell midden. Distance from known sites is ranked two. The site location is weighted to three or moderately high archaeological potential.

8.4.5. North Pruth Bay (10,608 cal BP)

North Pruth Bay (EjTa-15) is a site occupied multiple times through the Holocene on Calvert Island. Initial the site was recorded in 2012 as containing some chipped stone and two canoe runs in the intertidal zone. Excavations in 2013 identified several distinct cultural strata. One of the lower strata contained sixty-nine obsidian, basalt, chert, and andesite tools and flakes (McLaren et al. 2015). The earliest deposits date to 10,562 to 10,653 cal BP (9370±25 ^{14}C years (UCAMIS-128290)). There are water rolled flakes suggesting an older cultural deposit (McLaren et al. 2014). This site was then sporadically occupied through to the last hundred years.

North Pruth Bay's location is determined from published maps. The slope is zero degrees, which is a rank of five. The aspect is 73 degrees or east-northeast, which is a rank of one. The site is on a stream or has a distance of zero meters, which is a rank of four. There is a lake within 500 m of the site, which is a rank of five. The coast is approximately 20 m away, which is a rank of five. The nearby coastal sinuosity has a value of one, which is medium archaeological potential. There is a tributary junction approximately 20 m away, which is a rank of five. The nearest site is just over one km away, which is a rank of one. The overall weighted value for the site is three or moderately high archaeological potential.

8.4.6. North Kwakshua Channel (13,300 cal BP)

North Kwakshua Channel (EjTa-4) or Meay Channel I site was covered by the international popular media in 2018 for the identification of 13,000 cal BP footprints on the beach (PLOS 2018). The site is located on Calvert Island, west of Namu and north of Vancouver Island. Excavations by the University of Northern British Columbia field school extended 4.7 m into sediments and recovered stone, bone, antler, shell, and chipped stone objects. At approximately four meters deep, a shell layer was encountered that contained wet preserved wood that dates to approximately 6000 cal BP (McLaren et al. 2015). During testing of the intertidal zone in 2014, a suspected human track was found 60 cm below the beach surface. At the base of the depression, pieces of wood were recovered which were radiocarbon dated to between 13,095 to 13,169 cal BP (11,295± 30 ^{14}C years (UCIAMS-142561)) and 13,241 to 13,317 cal BP (11,435±30 ^{14}C years (UIAMS 149779)). In 2015 and 2016, an area of four by two meters was excavated to expose 28 human tracks oriented in different directions (McLaren et al. 2018).

The site location was provided with BC archaeology branch data. The slope is zero degrees, which is a rank of five. The aspect is 171 degrees or south, which is a rank of two. There is a stream approximately 350 m from the site, which is a rank of five. The closest lake is approximately 680 m away, which is a rank of three. The coast is adjacent or less than 10 m away, which is a rank of five. The sinuosity of the nearby coast is one, which is medium archaeological potential. The closest tributary junction is almost 500 m away, which is a rank of four. The closest site is almost 1.5 km away, which is a rank of one. The overall weighted value for North Kwakshua Channel is three or moderately high archaeological potential.

8.5. Site Summary

These eighteen sites provide a framework of the type of sites that might be encountered on the continental shelf of southeast Alaska. Raised marine terraces are producing newly identified sites associated with different sea levels at a time immediately postdating Pleistocene sea level rise. Similar terraces and caves on the continental shelf may also contain archaeological sites. These sites also provide a background for the types of artefacts that should be encountered. Although microblades may not be present at all of these early sites, the presence of unifacial choppers and bifacial foliate points extends further back into time. The sites reviewed here are in accord with Dixon's (2013a) Pebble Tool tradition with the Northwest Coast Microblade tradition in the northern sites and Pebble Tool tradition in the sites further south.

Of the eighteen sites, there are only three locations that are not accessed here. Of the other fifteen sites, one site was high potential, Kildidit Narrows (ElTa-18). All the remaining sites are moderately high archaeological potential. This indicates that the model will locate archaeological sites dating to before 10,000 cal BP when further testing is attempted.

9

Discussion

9.1. Introduction

Archaeological site location modelling is an iterative process. This research presents the fourth iteration of the model. The first iteration was a limited model focused on the edge of the continental shelf (-165 m) inferred to be the approximate LGM shoreline. It utilised modern hydrology and bathymetric features and the deeper areas were added by hand only in the study area. The second iteration investigated the region in 500 ^{14}C year time-slices based on an earlier sea level reconstruction. As the research progressed, issues were identified with defining tributaries, the importance of coastal sinuosity, defining probable palaeo-lakes, and the sea level reconstruction itself. Refinement of the sea level reconstruction resulted in the realisation that the -70 m contour interval was actually the land-sea interface 12,800 cal BP rather than 10,300 cal BP (9997 ^{14}C years) as originally interpreted. The new sea level reconstruction is more realistic as evident from the old curve having a 70 m sea level change in 500 ^{14}C years. The latest iteration is at a higher resolution (two-meters) for a smaller area and complements the third iteration of the study region.

9.2. Model

9.2.1. Resolution

Ideally, the DEM would be created at 50 cm resolution and the model at one-meter resolution, but this is limited by the computing power available and the resolution of the inputs to the DEM. This would mean that small archaeological features would be visible in the model. Unfortunately, the bathymetry and DEM for the region is not of sufficient resolution to produce a sub-meter continuous DEM. The one-meter resolution for the study area already incorporates extrapolation between points to generate, and thus it likely contains errors. However, it is anticipated that greater resolution will be possible in the future and incorporated in future iterations of this model, as has been the case with the 10-meter to two-meter iteration. This iteration of the model focuses on locations where resources were available for past people using environmental factors such as distance from freshwater and the amount of coastline and hence marine resource within the vicinity. These features are quantifiable at a 10-meter resolution and a two-meter resolution; thus, the model is appropriate for this analysis.

Multibeam data collection is the means to solve this problem and can be used to collect higher resolution bathymetric data. LiDAR imagery and metadata are also available for small areas of the Tongass National Forest. Combining high-resolution multibeam and LiDAR data would achieve sub-one-meter resolution for DEM development and facilitate construction of a higher resolution model. This is the process employed by the NWC modelling procedures described by Mackie, Fedje, and McLaren's (2018) model. The sub-meter resolution makes the location of archaeological sites on the landscape significantly easier and is a goal of this research.

Another issue of resolution is demonstrated by the geophysical survey results. The 2012 survey collected multibeam data at a coarse resolution. Consequently, post-processing was not able to generate one meter resolution modelling of the seafloor, a 1.5 m resolution version was the best result and it includes holes or data gaps. This results in making smaller landform changes or human-made depressions more difficult to discern in the multibeam data. High-resolution multibeam data was collected in subsequent years of the Gateway to the Americas II project, but these surveys were in other areas of the study region.

9.2.2. Moderately High vs. High Potential

This analysis has focused on model weights of three and four, or moderately high archaeological potential and high archaeological potential. Though having explained in chapter five that this was determined based on the distribution of the results, Kvamme's gain from chapter six (table 6-1) and the location analyses conducted in chapter eight make it evident that the weighted value of three is high archaeological potential. The 'moderately' qualifier is only added because weights above three exist. Of the fifteen site locations analysed in chapter eight, fourteen were located at an overall weight of three, with the last located at a weight of four. Considering the total area of weight four or high archaeological potential, discussed in chapter six, is 620 m² for the study area deep model and 540 m² for the study area 11,000 cal BP model, it is prudent to get boots-on-the-ground to tests these locations.

When analysing Kvamme's gain, the random locations in high archaeological potential areas of the study region models appear anomalous. When incorporated into the non-site landscape analysis, the 'high-potential' random locations are statistically expected. If one in one-hundred random points are supposed to actually be an archaeological site and there is only a small area that is ranked high-potential, then high potential areas of the study region models needs to be investigated too. There is 28.9 km² of high archaeological potential in the study region deep and 8.3 km² in the modern study area of high archaeological potential area. These are still reasonable areas to search, especially when they are limited to areas above modern sea level like the modern study area.

9.2.3. Spatial Statistics

Chapter six focuses on statistically testing or validating the high archaeological potential model. Kvamme's gain was extremely productive statistical test and aided in improving the model through multiple iterations. The other statistical tests were recommended to the author to test predictive models. This is the first iteration of the model that has been able to run Getis-Ord's General-G for high/low clustering due to the high computational demand of the process. Moran's I for autocorrelation was run for the last iteration of the model. These statistical tests did not seem useful to the overall analysis. If the assumption was that the environment variables were not spatially clusters and thus the model was not clustered, then this and other similar statistics could have been useful. This, however, contradicts much of the anthropological and archaeological literature about how people use space. We live in groups. Humans cluster around water sources and key resources. This is how predictive modelling is useful for archaeology; archaeological sites will be located 'near' the key resource within an area. The null hypotheses for these tests were rejected before the statistics were generated.

As the statistics were calculated based on the weighted value and there was limited-small areas covered by weight four or high archaeological potential, many of the results were invalidated by the lack of neighbours. Though recommend in the literature and by colleagues to test the model, these statistical tests would be better used to characterise the model input variables when they are not expected to be naturally correlated such as coastlines, river, and tributaries. This author does not recommend researchers use these test to validate archaeological predictive models in the future, unless indicated by the anthropological and archaeological literature of their study area.

9.3. Shakan Bay Glacial History

The multibeam and ROV survey of Shakan Bay identified new information about the geology for the bay and region. Glacial grooves and lineations are present on rocks identifying the direction of glacial movement and suggesting this area was overrun by glacial ice during the LGM or earlier. Based on the striations identified on the multibeam, glacial ice entered Shakan Bay from three different directions, one from the north, one from the southeast, and one from the south. These directions fit with glacial advances from the surrounding valley glaciers. Unfortunately, it is difficult to determine the sequence and timing of these advances. However, the extent of the striations and scour is limited and may indicate that parts of Shakan Bay may have been ice-free during the LGM.

Although not strictly archaeological in nature, these data are an important contribution to the growing body of knowledge defining Late Wisconsin and early Holocene palaeoecology of the Alexander Archipelago. These new data may contribute to the refinement of LGM glaciation that suggests this area was completely ice-covered during the LGM (Carrara et al. 2003). The multibeam geologic information is an important contribution to interpreting the glacial history of the United States Tongass National Forest and will be instrumental in future glacial reconstructions for the region and specifically for Shakan Bay.

9.4. Why Were No Sites Located?

Technically, one site was located. PET-714 is the location of the Restless, a ship that sank in 1910. The ship was a yawl vessel type. It was lost off Seven Mile Point between Shipley and Shakan Bays. The site was identified in the side scan sonar data. As we did not confirm the location using the ROV, we did not submit the site to the Alaska SHPO; they created the site from our final report (Dixon and Monteleone 2011).

Regarding prehistoric sites: the short answer, we did not do enough subsurface testing or high enough quality geophysical survey. If you compare the total amount of subsurface testing done during the Gateway to the Americas I and II project, in the entire study region, it is less than 40 m^3 for an area over 45,000 km^2. The probability of finding an archaeological site on land older than 10,000 cal BP is very slim due to taphonomic factors. With the limitations in subsurface testing during the Gateway to the Americas projects, which improved significantly in the final year, there was only a small chance we would find a site with the time allowed.

Shakan Bay was one of the first survey locations in 2010 and 2012. Though there were four years of fieldwork with the Gateway to the Americas project, Shakan Bay was a small part of the overall project. In 2010, only one day was spent surveying in Shakan Bay. In 2012, four days were spent with two vessels; one conducting the geophysical survey and one conducting sub-surface sampling with the van veen grab sampler. As the seasons progressed, the methods and results improved. The next side scan sonar survey did include overlapping coverage of the entire survey area defined. The next survey area in 2012 had the multibeam and sub-bottom profiler data conducted simultaneously. This saved time. Problems with tidal correction for the multibeam and file formats for the sub-bottom sonar were resolved as we moved into other survey areas. After the difficulties with sampling in 2012, a suction-dredge was developed and used. In this case, the suction dredge actually took two more field season to 'perfect'

Shakan Bay, the study area, was never tested with the suction dredge developed for the project. Testing locations were not always determined by the predictive model throughout the projects. In Shakan Bay, 73 successful (not empty) samples were collected; however, terrestrial sediments were never reached below the surface marine sediments. The grab sampler 'dropped' on a different spot each time and only penetrated a few centimetres into the sea floor. We learned a lot from time spent in Shakan Bay, but it was poorly sampled. . It is expected the next field season including ship-time to test subsurface deposits will be significantly more productive and will locate archaeological sites. More work is necessitated. However, funding needs to be secured for this research.

Another question to ask is: what did we think we could find? In an idealised world, a canoe with wood dating to older than 10,000 cal BP that sunk with freshly butchered fauna and some tools, including organic tools. The preservation environment possible underwater makes that not completely a dream. Though we know this is unlike; we would be contented to find some stone tools. Something similar to the flake found at Werner Bay (Fedje 2000; Fedje, Christensen, et al. 2005). A utilised flake with a barnacle post-deposition that can be dated for a limiting date.

One suggestion, after a conference presentation of this research, was coring to determine the sediment thickness. This was dismissed at the time due to the cost of the equipment in a remote location like southeast Alaska. It would have to be done as part of a larger project; not part of Gateway to the Americas. If the opportunity arose, the sub-bottom data would be invaluable in identifying specific locations to take cores within the study region. In Shakan Bay, if cores could be collected in the depressions that are thought to have been a series of estuaries or connected lakes at lower sea levels, they would have the potential to provide more than just the sea level history for the bay. It is anticipated there are intact palaeosols and pollen at the bottom of these depressions. Cores could provide information about the glacial refugia timing and revegetation in a medium that could be radiocarbon dated.

Soil probes and percussion coring are becoming the new standards for archaeological survey on the northern NWC (A. Cannon 2000; R. J. Carlson and Baichtal 2015; Letham et al. 2016; Martindale and Supernant 2009; McLaren et al. 2007). One proposal is sampling in the intertidal and underwater in the study area, similar to the work done at around Namu, Dundas Island, and around Prince Rupert. Using recreational diving limits of approximately 30 m and the current sea level reconstruction, this research could fill in the sea level reconstruction back to almost 12,000 cal BP. Though labour intensive, these methods are locating sites with minimal logistical costs on the BC mainland (Letham et al. 2015, 2016).

9.5. Beringia

The now submerged Beringian continent (figure 9-1) is a vast area that likely holds the key to the peopling of the Americas. Research about Beringia is focused on the eastern and western edges, which are Alaska and eastern Russia (Dixon and Monteleone 2014; Wygal 2018; G. M. MacDonald et al. 2012; Hoffecker et al. 2016; Hoffecker, Elias, and Rourke 2014; Goebel and Slobodin 1999; Raghavan et al. 2015). The Beringian continent was land during the LGM. Parts of it were open grasslands, especially to the south with some shrub-brush forests, and arctic tundra to the north. This was not a desolate wasteland; it would have been teaming with flora and fauna including now-extinct megafauna. People hunting and gathering in Siberia would not have known they were even in a new continent when they finally migrated to Alaska. Some would have followed game; others would have moved for demographic pressures. Beringia was large enough that assuming a single group or tradition occupied the landscape is impractical. Those that later arrived in what is now central and northern Alaska following a terrestrial or Arctic Ocean route could have been a very different group than those that followed the kelp highway along the southern coast of Beringia southward through the NWC and into lower American mainland, south of the glaciers.

Using the kelp highway theory, the high archaeological potential model for southeast Alaska south should be applicable almost anywhere within the kelp highway. Model weights would need to be adjusted based on the local archaeological site data. At a minimum, the model presented and tested in this report should work in a similar environment, such as an island archipelago with similar resources.

In 2013 for the Paleoamerican Odyssey conference in Santa Fe, Andrew Wickert and the author presented a poster of a variation of this model for the submerged southern Beringian Archipelago (Monteleone and Wickert 2013). Alongside this was a poster by Wickert, Monteleone, Mitrovia, and Anderson (2013) reconstructing the palaeogeography of Beringia. The palaeogeography incorporated an updated and nonuniform sea level change across Beringia and using the TraCE-21K palaeoclimate general circulation model to asses past temperatures, precipitation, and evapotranspiration across Beringia. This palaeoclimatic model generated the streams and coastline for the archaeological predictive model.

These Beringian models were very coarse, though they are at a higher resolution than others in the literature. The DEM was at a resolution of 677 m. The model was generated at 900 m resolution in 1000 cal BP year intervals over the southeast portion of the Bering Sea from 10,000 cal BP to 18,000 cal BP. This is the area includes St. Mathews Island, Nunivak, and the Pribilof Islands (figure 9-1). The weighted overlay values had to be modified as there were no known archaeological sites to incorporate into the model. The authors did not have access to the site data for this quick-desktop project. Overall, there was not a lot of high potential at 10,000 cal BP, except near Nunivak. The 18,000 cal BP reconstruction showed much larger areas of high potential expanding away from the current islands.

Again, the area of this project was 419,000 km^2. As we did not have archaeological site data, Kvamme's gain was not calculated. The models did not produce a value

Uncovering Submerged Landscapes

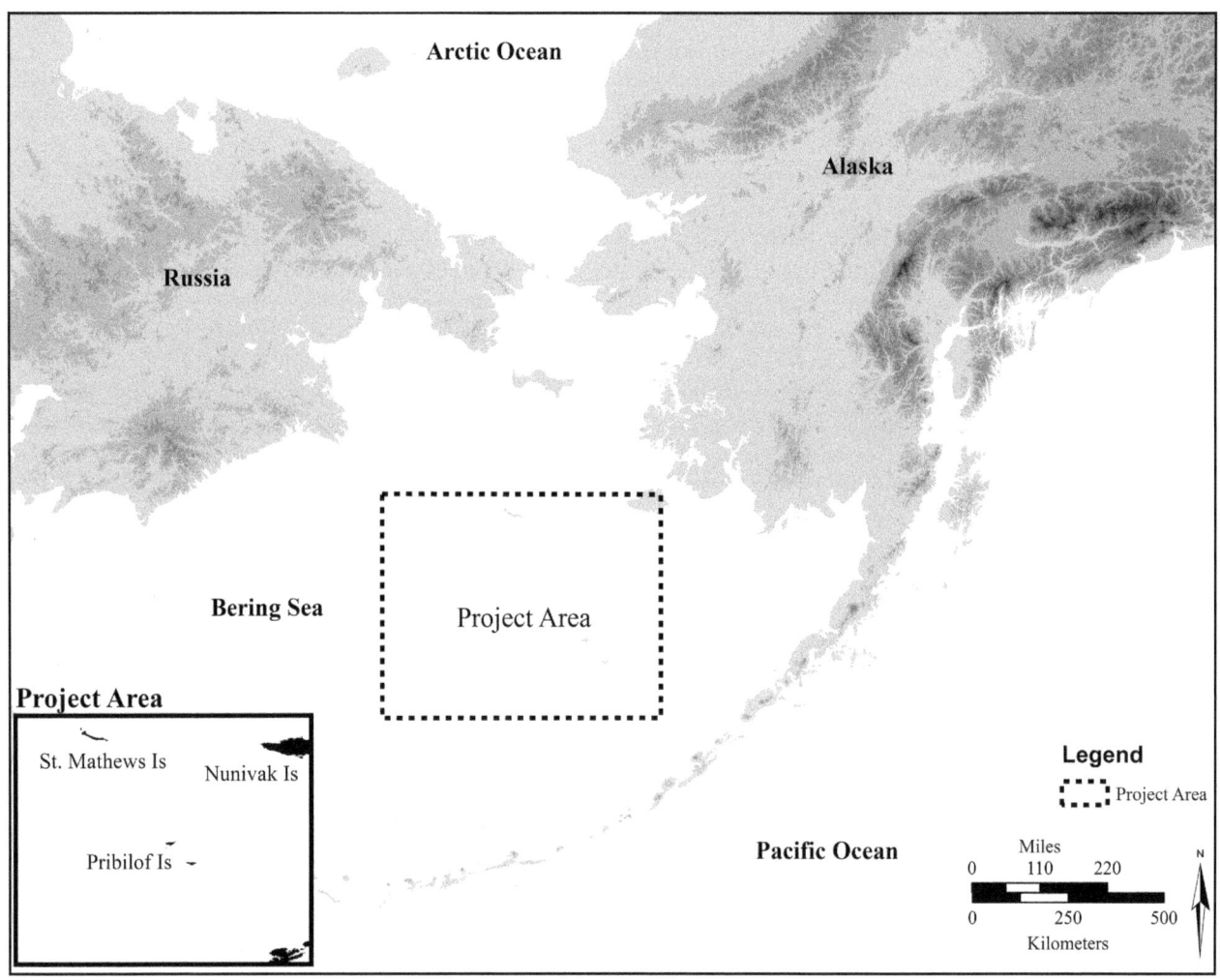

Figure 9-1. Map of Bering Sea with inset map of southern Beringian archipelago.

of four or high archaeological potential, only moderately high archaeological potential. The 10,000 cal BP model had only three per cent of the total area of the model at moderately high archaeological potential. Additionally, the area of the model was only three per cent of the total project area due to lower sea levels. By 18,000 cal BP, the model area was 24 per cent of the project area and the moderately high archaeological potential area was only two per cent.

The southern archipelago was selected for this analysis, as it would have shared some aspects with the Alexander Archipelago in southeast Alaska. At lower sea levels both were sheltered areas that would have provided some protection from the full fetch of the open ocean. Both areas are near larger mainland areas. As the model was developed based partially on the archaeological site data from southeast Alaska, a similar geographic environment was sought. If a straight section of coastline were selected, the model would have been less useful.

This side project still needs refinement and additional iterations before publication, including incorporating known archaeological sites and adding other recent environmental data from Beringia (and other data shared with the authors after the conference presentation).

10

Conclusion

This research cannot yet accept or reject the hypothesis that the archaeological record of Southeast Alaska extends to areas of the continental shelf that were submerged by post-Pleistocene sea level rise beginning around 10,600 cal BP (9,400 ^{14}C years) due to the limited field testing conducted. This is a huge area and the high potential model developed for this research has only been tested in a very limited area. Tests of the model, thus far, have indicated that the model is useful in predicting site locations. Of the fifteen pre-10,000 cal BP sites that could be tested against the model all would have been predicted by the four models tested. This model is specific to sea level and the NWC culture area. It has the ability to be adapted to other regions by changing the sea level reconstructions and reassessing the weights for the input model to accommodate the culture history of the new region.

This research has demonstrated the cost-effectiveness and benefits of using a land-use model to identify high potential areas for archaeological sites before an underwater archaeological survey. With the expense of underwater archaeology, the methods utilised will be useful for future large-scale underwater archaeological surveys. The model generated for this research has proven successful when tested against known and new terrestrial archaeological sites. New underwater archaeological sites will be located with further testing.

The Gateway to the Americas projects has completed. The author is actively applying for funding to continue testing this high archaeological potential model. One avenue is to conduct auger testing similar to what has been done along the mainland of BC in the intertidal and foreshore (A. Cannon 2000; R. J. Carlson and Baichtal 2015; Letham et al. 2016; Martindale and Supernant 2009; McLaren et al. 2007). In Shakan Bay, we would be able to investigate to approximately the 12,000 cal BP shoreline using divers to conduct auger tests. Essentially, this research has not ended. New approaches and funding are constantly being explored to continue to locate submerged sites in southeast Alaska.

One of the intentions behind this research is to develop a series of methods that work in shallower (less than 50 meters) water that could be applied in deep/open water on the outer coast (down to 165 meters on the outer coast) of southeast Alaska and in Beringia. Once sites have been located using the methods employed by the Gateway to the Americas and/or refined versions of these methods/models, future proposals will explore using these methods in the Bering Sea. This would then have implications not just for how people first migrated to the Americas, but also the Beringian Standstill hypothesis (Moreno-Mayar et al. 2018; Hoffecker et al. 2016; Faught 2017).

This research significantly advances knowledge of NWC archaeology by surveying an archaeologically unexplored region: the continental shelf of southeast Alaska. As off-shore wind and oil exploration intensify in North America, the methods developed in this proposal could aid cultural resource management (CRM) in protecting North American submerged cultural heritage. The Bureau of Ocean Energy Management (BOEM) in Alaska has already recognised the value of this research in advance of oil exploration in the Beaufort and Chukchi Seas. Understanding the human response to sea level change is also relevant to the anthropogenic acceleration of climate change and the associated rise in global sea level. Finally, locating early archaeological sites on the continental shelf of southeast Alaska has the potential to contribute to one of the oldest anthropological questions: the colonisation of the American continents.

Bibliography

Ackerman, Robert E. 1968. *The Archaeology of the Glacier Bay Region, Southeastern Alaska*. Pullman: Washington State University, Report of Investigations No 44.

———. 1996. 'Ground Hog Bay, Site 2.' In *American Beginnings: The Prehistory and Palaeoecology of Beringia*, edited by Frederick Hadleigh. West and Constance F. West, 424–30. Chicago: University of Chicago Press.

Ackerman, Robert E., Thomas D. Hamilton, and Robert Stuckenrath. 1979. 'Early Culture Complexes of the Northern Northwest Coast.' Edited by Roy L. Carlson and Luke R. Dalla Bona. *Canadian Journal of Archaeology* 3. Vancouver: UBC Press:195–209.

Ager, Thomas A., and Joseph G. Rosenbaum. 2007. 'Late Glacial-Holocene Pollen-Based Vegetation History from Pass Lake, Prince of Wales Island, Southeastern Alaska.' *Studies by the U.S. Geological Survey in Alaska, 2007, Professional Paper 1760-G*, 19.

Ager, Thomas A., Paul E. Carrara, Jane L. Smith, Victoria Anne, and Joni Johnson. 2010. 'Postglacial Vegetation History of Mitkof Island, Alexander Archipelago, Southeastern Alaska.' *Quaternary Research* 73 (2):259–68. https://doi.org/10.1016/j.yqres.2009.12.005.

Ahler, Stanley A. 1989. 'Mass Analysis of Flaking Debris: Studying the Forest Rather Than the Tree.' In *Alternative Approaches to Lithic Analysis*, edited by Donald O. Henry and George H. Odell, 85–118. Washington, D.C.: Archaeological papers of the American Anthropological Association, no 1.

Altschul, Jeffrey A. 1988. 'Models and the Modeling Process.' In *Quantifying the Present and Predicting the Past: Theory, Method, and Application of Archaeological Predictive Modeling*, edited by W. James Judge and Lynne Sebastian, 61–96. Denver: US Department of the Interior, Bureau of Land Management.

Altschul, Jeffrey A., and Christopher L. Nagle. 1988. 'Collecting New Data for the Purpose of Model Development.' In *Quantifying the Present and Predicting the Past: Theory, Method, and Application of Archaeological Predictive Modeling*, edited by W. James Judge and Lynne Sebastian, 257–99. Denver: US Department of the Interior, Bureau of Land Management.

Altschul, Jeffrey A., Lynne Sebastian, and K. Heidelberg. 2004. *Predicitive Modeling in the Military. Similar Goals, Divergent Paths*. Rio Rancho: SRI Foundation.

Ames, D. P., J. S. Horsburgh, Y. Cao, J. Kadlec, T. Whiteaker, and D. Valentine. 2012. 'HydroDesktop: Web Services-Based Software for Hydrologic Data Discovery, Download, Visualization, and Analysis.' Environmental Modelling & Software. Vol 37. 2012. http://dx.doi.org/10.1016/j.envsoft.2012.03.013.

Ames, Kenneth M., and Herbert D. G. Maschner. 1999. *Peoples of the Northwest Coast: Their Archaeology and Prehistory*. New York: Thames and Hudson.

Anderson, David G., Thaddeus G. Bissett, and Stephen J. Yerka. 2013. 'The Late-Pleistocene Human Settlement of Interior North America: The Role of Physiography and Sea-Level Change.' In *Paleoamerican Odyssey*, edited by Kelly E. Graf, Caroline V. Ketron, and Michael R. Waters, 183–203. Bryan: Center for the Study of the First Americas, Department of Anthropology, Texas A&M University.

Anderson, R. G. 1991. 'Contributions to the Geology and Geophysics of Northwestern British Columbia and Southeastern Alaska.' *Canadian Journal of Earth Sciences* 28:835–39.

Anschuetz, Kurt F., Richard H. Wilshusen, and Cherie L. Scheick. 2001. 'An Archaeology of Landscapes: Perspectives and Directions.' *Journal of Archaeological Research* 9 (2):157–211. https://doi.org/10.1023/A:1016621326415.

Antevs, Ernst. 1929. 'Maps of the Pleistocene Glaciations.' *Bulletin of the Geological Society of America* 40:631–720.

Arnold, Thomas G. 2006. 'The Ice-Free Corridor: Biogeographical Highway or Environmental Cul-de-Sac.' Simon Fraser University.

Baichtal, James F., and Risa J. Carlson. 2010. 'Development of a Model to Predict the Location of Early Holocene Habitation Sites along the Western Coast of Prince of Wales Island and the Outer Islands, Southeast Alaska.' In *2010 GSA Denver Annual Meeting*.

Baichtal, James F., G. Streveler, and Terrence Fifield. 1997. 'The Geological, Glacial and Cultural History of Southern Southeast.' *Alaska Geographic* 24 (1):6–31.

Bailey, Geoffrey N., and Nicholas C. Flemming. 2008. 'Archaeology of the Continental Shelf: Marine Resources, Submerged Landscapes and Underwater Archaeology.' *Quaternary Science Reviews* 27 (23–24):2153–65. https://doi.org/10.1016/j.quascirev.2008.08.012.

Ballard, Robert D. 2008. 'Introduction.' In *Archaeological Oceanography*, edited by Robert D. Ballard, ix–x. Princeton, NJ: Princeton University Press.

Balter, Michael. 2008. 'Archaeology: Ancient Algae Suggest Sea Route for First Americans.' *Science* 320 (5877):729. https://doi.org/10.1126/science.320.5877.729.

Barber, Russel, and Michael E. Roberts 1979. *Archaeology and Paleontology. A Summary and Analysis of Cultural Resource Information on the Continental Shelf from the Bay of Fundy to Cape Hatteras, Final Report, Vol II*. Cambridge: Institute for Conservation Archaeology Report 88. Peabody Museum, Harvard University.

Barrie, J. Vaughn, and Kim W. Conway. 1999. 'Late Quaternary Glaciation and Postglacial Stratigraphy of the Northern Pacific Margin of Canada.' *Quaternary Research* 51 (2):113–23. https://doi.org/10.1006/qres.1998.2021.

Barron, John A., David Bukry, Walter E. Dean, Jason A. Addison, and Bruce P. Finney. 2009. 'Paleoceanography of the Gulf of Alaska during the Past 15,000 Years: Results from Diatoms, Silicoflagellates, and Geochemistry.' *Marine Micropaleontology* 72 (3–4):176–95. https://doi.org/10.1016/j.marmicro.2009.04.006.

Baucaire, Rene. 1998. 'The Fos Underwater Excavation. Bulletin de La Societe Des Amis Du Vielis Istres. Tome Quatorzieme, La Pensee Universitaire, Aix-En-Provence. Translated by Thomas A. Babits. Reprinted.' In *Maritime Archaeology: A Reader of Substantive and Theoretical Contributions*, edited by Babits and Tilburg, 9–15. New York: Plenum Press.

Bell, James A. 1994. *Reconstructing Prehistory: Scientific Methods in Archaeology*. Philadelphia: Temple University Press.

Bender, Barbara. 2002. 'Time and Landscape.' *Current Anthropology* 43 (S4):S103–12. https://doi.org/10.1086/339561.

Benjamin, Jonathan. 2010. 'Submerged Prehistoric Landscapes and Underwater Site Discovery: Reevaluating the 'Danish Model' for International Practice.' *Journal of Island and Coastal Archaeology* 5 (2):253–70. https://doi.org/10.1080/15564894.2010.506623.

Bettinger, Robert L., Raven Garvey, and Shannon Tushingham. 2015. *Hunter-Gatherers: Archaeological and Evolutionary Theory, Second Edition*. 2nd ed. New York: Springer.

Bevan, A., and James Conolly. 2005. 'Mutliscalar Approaches to Settlement Pattern Analysis.' In *Confronting Scale in Archaeology: Issues of Theory and Practice*, edited by Gary Lock and B. Molyneaux, 217–34. New York: Kluwer Academic Publishers.

Binford, Lewis R. 1977. 'General Introduction.' In *For Theory Building in Archaeology*, edited by Lewis R. Binford, 1–10. New York: Academic Press. https://doi.org/10.1017/CBO9781139020503.002.

———. 1980. 'Willow Smoke and Dogs' Tails: Hunter-Gatherer Settlement Systems and Archaeological Site Formation.' *American Antiquity* 45 (1):4–20. https://doi.org/10.2307/279653.

———. 1981. 'Relics to Artifacts and Monuments to Assemblages: Changing Conceptual Frameworks.' In *Bones: Ancient Men and Modern Myths*, edited by Lewis R. Binford, 3–20. New York: Academic Press.

Bird, Douglas W., and James F. O'Connell. 2006. 'Behavioral Ecology and Archaeology.' *Journal of Archaeological Research* 14 (2):143–88. https://doi.org/10.1007/s10814-006-9003-6.

Bird, Douglas, and Brian Codding. 2008. 'Human Behavioral Ecology and the Use of Ancient Landscapes.' In *Handbook of Landscape Archaeology*, edited by Bruno David and Julian Thomas, 396–408. Walnut Creek, CA, CA: Left Coast Press.

Bjerck, Hein Bjartmann. 1995. 'The North Sea Continent and the Pioneer Settlement of Norway.' In *Man and Sea in the Mesolithic: Coastal Settlement above and below Present Sea Level*, edited by Anders Fischer, Proceeding, 131–44. Oxford: Oxbow Monograph 53.

———. 2009. 'Colonizing Seascapes: Comparative Perspectives on the Development of Maritime Relations in Scandinavia and Patagonia.' *Arctic Anthropology* 46 (1–2):118–31. https://doi.org/10.1353/arc.0.0019.

Blackman, Margaret B. 1990. 'Haida: Traditional Culture.' In *Handbook of North American Indians, Volume 7: Northwest Coast*, edited by Wayne Suttles and William C. Sturtevant, 240–60. Washington, D.C.: Smithsonian Institution Press.

Blaise, Bertrand, John J. Clague, and Rolf W. Mathewes. 1990. 'Time of Maximum Late Wisconsin Glaciation, West Coast of Canada.' *Quaternary Research* 34:282–95.

Boas, Franz. 1935. *Kwakiutl Culture: As Reflected in Mythology*. New York: American Folk-Lore Society.

———. 1982. '[1940] The Social Organization of the Tribes of the North Pacific Coast.' In *Race, Language, and Culture*, 370–78. Chicago: The University of Chicago Press.

Boaz, J., and E. Uleberg. 2000. 'Quantifying the Non-Quantifiable: Studying Hunter-Gatherer Landscape.' In *Beyond the Map: Archaeology and Spatial Technologies*, edited by Gary Lock, 101–15. Amsterdam: IOS Press.

Borden, Charles E. 1950. 'Preliminary Report on Archaeological Investigations in the Fraser Delta Region.' *Anthropology in British Columbia* 1:13–26.

———. 1951. 'Facts and Problems of Northwest Coast Prehistory.' *Anthropology in British Columbia* 2:35–57.

———. 1975. *Origins and Development of Early Northwest Culture to about 3000 BC*. Ottawa: National Museum of Man, Mercury Series, Archaeological Survey of Canada Paper 45.

Bowens, Amanda. 2009. *Underwater Archaeology: The NAS Guide to Principles and Practice*. Oxford: Blackwell Publishers.

Bower, Patrick W. 1994. 'Heritage Resource Investigations for the Northwest Baranof Project.' Sitka.

Braje, Todd J., Tom D. Dillehay, Jon M. Erlandson, Richard G. Klein, and Torben C. Rick. 2017. 'Finding the First Americans: The First Humans to Reach the Americas Are Likely to Have Come via a Coastal Route.' *Science* 358 (6363). https://doi.org/10.1126/science.aao5473.

Breen, Colin, and Paul Lane. 2004. 'Archaeological Approaches to East Africa's Changing Seascapes.' *World Archaeology* 35 (3):469–89. https://doi.org/10.1080/0043824042000185838.

Brew, David A., Susan M. Karl, David F. Barnes, Robert C. Jachens, Arthur B. Ford, and Robert Horner. 1991. 'A Northern Cordilleran Ocean-Continent Transect: Sitka Sound, Alaska, to Atlin Lake, British Columbia.' *Canadian Journal of Earth Sciences* 28 (840–853).

Brown, James H. 1995. *Macroecology*. Chicago: The University of Chicago Press.

Bryn, Petter, Marek E. Jasinski, and Fredrik Soreide. 2007. *Ormen Lange: Pipelines and Shipwrecks*. Oslo: Universitetsforiaget.

Cannon, Aubrey. 1998. 'Contingency and Agency in the Growth of Northwest Coast Maritime Economies.' *Arctic Anthropology* 35 (1):57–67.

———. 2000. 'Settlement and Sea-Levels on the Central Coast of British Columbia: Evidence from Shell Midden Cores.' *American Anthropologist* 65 (1):67–77. https://doi.org/10.1525/aa.1961.63.5.02a00100.

Cannon, Aubrey, and Dongya Y. Yang. 2011. 'Pushing Limits and Finding Interpretive Balance: A Reply to Monks and Orchard.' *American Antiquity* 76 (3):585–95.

Cannon, Michael D., and Jack M. Broughton. 2010. 'Evolutionary Ecology and Archaeology.' In *Evolutionary Ecology and Archaeology: Applications to Problems in Human Evolution and Prehistory*, edited by Jack M. Broughton and Michael D. Cannon, 1–12. Salt Lake City: University of Utah Press.

Carleton, W.C., J. Conolly, and G. Ianonne. 2012. 'A Locally-Adaptive Model of Archaeological Potential (LAMAP).' *Journal of Archaeological Science* 39 (11). Elsevier Ltd:3371–85. https://doi.org/10.1016/j.jas.2012.05.022.

Carlson, Catherine. 1996. 'Diversity and Complexity in Prehistoric Maritime Societies - a Gulf of Maine Perspective.' *American Antiquity* 61 (3):621–22.

Carlson, Risa J. 2007. 'Current Models for the Human Colonization of the Americas: The Evidence from Southeast Alaska.' Cambridge University.

Carlson, Risa J., and James F. Baichtal. 2015. 'A Predictive Model for Locating Early Holocene Archaeological Sites Based on Raised Shell-Bearing Strata in Southeast Alaska, USA.' *Geoarchaeology* 30 (2):120–38. https://doi.org/10.1002/gea.21501.

Carlson, Roy L. 1979. 'The Early Period on the Central Coast of British Columbia.' *Canadian Journal of Archaeology* 3 (3):211–28.

———. 1990. 'Cultural Antecedents.' In *Handbook of North American Indians, Volume 7: Northwest Coast*, edited by Wayne Suttles and William C. Sturtevant, 60–69. Washington, D.C.: Smithsonian Institution Press.

———. 1993. 'Soil Science and a Northwest Coast Shell Midden (Review of Stein Ed).' *The Review of Archaeology* 14:15–21.

———. 1996. 'Early Namu.' In *Early Human Occupation in British Columbia*, edited by Roy L. Carlson and Luke Della Bona, 83–110. Vancouver: University of British Columbia Press.

———. 1998. 'Coastal British Columbia in the Light of North Pacific Maritime Adaptations.' *Arctic Anthropology* 35 (1):23–35.

———. 2000. 'Review of Peoples of the Northwest Coast: Their Archaeology and Prehistory by Kenneth M. Ames and Herbert D. G. Maschner. London: Thames and Hudson, 1999, 288 Pp.' *Journal of Archaeological Research* 56 (2):254–56.

Carrara, Paul E., Thomas A. Ager, and James F. Baichtal. 2007. 'Possible Refugia in the Alexander Archipelago of Southeastern Alaska during the Late Wisconsin Glaciation.' *Canadian Journal of Earth Sciences* 44:229–44. https://doi.org/10.1139/e06-081.

Carrara, Paul E., Thomas A. Ager, James F. Baichtal, and D. Paco VanSistine. 2003. 'Map of Glacial Limits and Possible Refugia in the Southern Alexander Archipelago, Alaska, During the Late Wisconsin Glaciation.' *United States Geologic Survey, Miscellaneous Field Studies Map* MF-2424.

Casey, Edward S. 2008. 'Place in Landscape Archaeology: A Philosophical Prelude.' In *Handbook of Landscape Archaeology*, edited by Bruno David and Julian Thomas, 44–50. Walnut Creek, CA, CA: Left Coast Press.

Chapman, Henry P., and Philip R. Chapman. 2005. 'Seascapes and Landscapes - The Siting of the Ferriby Boat Finds in the Context of Prehistoric Pilotage.' *International Journal of Nautical Archaeology* 34 (1):43–50. https://doi.org/10.1111/j.1095-9270.2005.00042.x.

Ch'ng, Eugene, Henry Chapman, Vincent Gaffney, Phil Murgatroyd, Chris Gaffney, and Wolfgang Neubauer.

2011. 'From Sites to Landscapes: How Computing Technology Is Shaping Archaeological Practice.' *Computer* 44 (7):40–46. https://doi.org/10.1109/MC.2011.162.

Christensen, Arne Emil. 1997. 'Review of Archaeology under Water: Research and Reports on Underwater Archaeology in German and Swiss Lake Dwellings and the North Sea - Schilchtherle, H.' *International Journal of Nautical Archaeology* 26 (3):268–69.

Clague, Jon H., Rolf W. Mathewes, and Thomas A. Ager. 2004. 'Environments of Northwestern North America before the Last Glacial Maximum.' In *Entering America : Northeast Asia and Beringia before the Last Glacial Maximum*, edited by David B. Madsen, 63–94. Salt Lake City: University of Utah Press.

Clark, David L. 1968. *Analytical Archaeology*. London: Methuen.

Cliquet, Dominque, Sylvie Coutard, Martine Clet, Jean Allix, Bernadette Tessier, Frank Lelong, Anges Baltzer, et al. 2011. 'The Middle Palaeolithic Underwater Site of La Mondree, Normandy, France.' In *Submerged Prehistory*, edited by Jonathan Benjamin, Clive Bonsall, Catriona Pickard, and Anders Fischer, 111–28. Oxford: Oxbow Books.

Coastal Environments, Inc. (CEI). 1977. 'Cultural Resources Evaluation of the Northern Gulf of Mexico Continental Shelf, Volume II Historical Cultural Resources.' Washington, D.C.

Coleman, Dwight F. 2008. 'Archaeology and Geological Oceanography of Undated Coastal Landscapes: An Introduction.' In *Archaeological Oceanography*, edited by Robert D. Ballard, 177–99. Princeton, NJ: Princeton University Press.

Coleman, Dwight F., and Robert D. Ballard. 2008. 'Oceanographic Methods for Underwater Archaeological Surveys.' In *Archaeological Oceanography*, edited by Robert D. Ballard, 3–14. Princeton, NJ: Princeton University Press.

Coleman, Dwight F., and Kevin McBride. 2008. 'Underwater Prehistoric Archaeological Potential of the Southern New England Continental Shelf off Block Island.' In *Archaeological Oceanography*, edited by Robert D. Ballard, 200–223. Princeton, NJ: Princeton University Press.

Coles, Bryony J. 1998. 'Doggerland: A Speculative Survey.' *Proceedings of the Prehistoric Society* 64 (January):45–81. https://doi.org/10.1017/S0079497X00002176.

———. 2000. 'Doggerland: The Cultural Dynamics of a Shifting Coastline.' *Coastal and Estuarine Environments: Sedimentology, Geomorphology and Geoarchaeology* 175 (1):393–401. https://doi.org/0305-8719.

———. 2004. 'The Development of Wetland Archaeology in Britain.' In *Living on the Lake in Prehistoric Europe: 150 Years of Lake-Dwelling Research*, edited by Francesco Menotti, 98–114. New York: Rutledge.

Collins, Michael B., Dennis J. Stanford, Darrin L. Lowery, and Bruce A. Bradley. 2013. 'North America before Clovis: Variance in Temporal/Spatial Cultural Patterns, 27,000 - 13,000 Cal Yr BP.' In *Paleoamerican Odyssey*, edited by Kelly E. Graf, Caroline V. Ketron, and Michael R. Waters, 521–40. Bryan: Center for the Study of the First Americans, Department of Anthropology, Texas A&M University.

Conrad, Clinton P. 2013. 'The Solid Earth's Influence on Sea Level.' *Bulletin of the Geological Society of America* 125 (7–8):1027–52. https://doi.org/10.1130/B30764.1.

Cook Hale, Jessica W., and Ervan G. Garrison. 2017. 'Spatial Statistical Analysis of Coastal Plain Paleoindian Site Distributions and Paleoecology: Implications for the Search for Offshore Submerged Sites in Georgia.' *Early Georgia: The Society for Georgia Archaeology* 45 (1 & 2):167–80.

Croome, Angela. 2004. 'Thoughts on Thirty Years as Reviews Editor.' *International Journal of Nautical Archaeology* 33 (2):352–53. https://doi.org/10.1111/j.1095-9270.2004.032.x.

Dalla Bona, Luke. 1994. 'Cultural Heritage Resource Predictive Modelling Project: Volume 4: A Predictive Model of Prehistoric Activity Location for Thunder Bay District, Ontario.'

———. 2003. 'Predictive Maps and Archaeological Conservation in Ontario.' In *Forschungen Zur Archäologie Im Land Brandenburg Band 8: Landschaftsarchäologie Und Geographische Informationssysteme. Prognosekarten, Besiedlungsdynamik Und Prähistorische Raumordn*, edited by Jurgen Kunow and Muller Johannes, 99–105. Abbildungen: Archaoprognose.

Dalla Bona, Luke, and Linda Larcombe. 1996. 'Modeling Prehistoric Land Use in Northern Ontario.' In *New Methods, Old Problems: Geographic Information Systems in Modern Archaeological Research*, edited by Herbert D. G. Maschner, 252–71. Carbondale: Occasional Paper No 23, Center for Archaeological Investigations, Southern Illinois University.

Daly, P. T., and Gary R. Lock. 1999. 'Timing Is Everything: Commentary on Managing Temporal Variables in Geographic Information Systems.' In *New Techniques for Old Times: CAA 98 Computer Applications and Quantitative Methods in Archaeology*, edited by J. A. Barcelo, I Briz, and Vila A. S., 287–93. Oxford: BAR Publishing, BAR International Series S757.

David, Bruno, and Julian Thomas. 2008. 'Landscape Archaeology: Introduction.' In *Handbook of Landscape Archaeology*, edited by Bruno David and Julian

Thomas, 27–43. Walnut Creek, CA: Left Coast Press Inc.

Davidson-Arnott, Robin. 2010. *Introduction to Coastal Processes and Geomorphology*. Cambridge: Cambridge University Press.

Davis, Stanley D. 1989. *The Hidden Falls Site, Baranof Island, Alaska*. Anchorage: Aurora, Alaska Anthropological Association Monograph Series V.

———. 1990. 'Prehistory of Southeastern Alaska.' In *Handbook of North American Indians, Volume 7: Northwest Coast*, edited by Wayne Suttles, 197–202. Washington, D.C.: Smithsonian Institution.

———. 1996. 'Hidden Falls.' In *American Beginnings: The Prehistory and Paleoecology of Beringia*, edited by F. W. West, 413–24. Chicago: University of Chicago.

Dawson, George M. 1894. 'Geological Notes on Some of the Coasts and Islands of the Bering Sea and Vicinity.' *Bulletin of the Geological Society of America* 5 (February):117–46.

Dean, M. Ferrari, Ian Oxley, M. Rednap, and K. Watson. 1992. *Archaeology Underwater: The NAS Guide to Principle and Principle*. London: Nautical Archaeology Society.

Deeben, Jos, Dann P. Hallewas, J. Kolen, and R. Wiemer. 1997. *Beyond the Crystal Ball: Predictive Modelling as a Tool in Archaeological Heritage Management and Occupation History*. Edited by W. J. H. Willems, H. Kars, and Dann P Hallewas. *Archaeological Heritage Management in the Netherland: Fifty Years State Service for Archaeological Investigations*. Amersfoort.

Demboski, John R., Karen D. Stone, and Joseph A. Cook. 1999. 'Further Perspectives on the Haida Gwaii Glacial Refugium.' *Evolution* 53 (6):2008–12.

Descartes, Rene. 1991. *[1647] Principles of Philosophy*. Translated by Valentine Rodger Miller. Netherlands: Kluwer Academic Publishers.

Dillehay, Tom D., C. Ramírez, M. Pino, M. B. Collins, J. Rossen, and J. D. Pino-Navarro. 2008. 'Monte Verde: Seaweed, Food, Medicine, and the Peopling of South America.' *Science* 320 (5877):784–86. https://doi.org/10.1126/science.1156533.

Dincauze, Dena F. 2000. *Environmental Archaeology: Principles and Practice*. Cambridge: Cambridge University Press.

Dixon, E. James. 1979. 'A Predictive Model for the Distribution of Archaeological Sites on the Bering Continental Shelf.' Brown University.

———. 1983. 'Pleistocene Proboscidean Fossils from the Alaskan Outer Continental Shelf.' *Quaternary Research* 20:113–19.

———. 1989. 'Reassessing the Occurrence and Survival of Submerged Terrestrial Archaeological Sites on the Bering Continental Shelf.' In *Underwater Archaeological Proceedings from the Society of Historical Archaeology Conference*, edited by J. Barto III Arnold. Baltimore.

———. 1993. *Quest for the Origins of the First Americans*. Albuquerque: University of New Mexico.

———. 1999. *Bones, Boats & Bison : Archeology and the First Colonization of Western North America*. Albuquerque: University of New Mexico Press.

———. 2011a. 'Arrows, Atlatls, and Cultural-Historical Conundrums.' In *From Yenisei to the Yukon*, edited by Ted Geobel and Ian Buvit, 362–69. College Station: Texas A&M University Press.

———. 2011b. 'The Pleistocene/Holocene Transition along the Northwest Coast.' In *Paper Presented at the 38th Annual Meeting of the Alaska Anthropology Association*. Fairbanks.

———. 2013a. *Arrows and Atls-Atls: A Guide to the Archeology of Beringia*. United States: Department of the Interior Shared Beringian Heritage Program (U.S.).

———. 2013b. 'Late Pleistocene Colonization of North America from Northeast Asia: New Insights from Large-Scale Paleogeographic Reconstructions.' *Quaternary International* 285. Elsevier Ltd and INQUA:57–67. https://doi.org/10.1007/978-3-319-15138-0_12.

Dixon, E. James, and Kelly Rose Bale Monteleone. 2011. 'Final Report for Gateway to the Americas (2010-02).' Anchorage.

———. 2014. 'Gateway to the Americas: Underwater Archeological Survey in Beringia and the North Pacific.' In *Prehistoric Archaeology on the Continental Shelf: A Global Review*, edited by Amanda Evans, Joseph C. Flatman, and Nicholas C. Flemming, 95–114. New York: Springer. https://doi.org/10.1007/978-1-4614-9635-9.

Dixon, E. James, Timothy H. Heaton, Terence E. Fifield, Thomas D. Hamilton, David E. Putnam, and Frederick Grady. 1997. 'Late Quaternary Regional Geoarchaeology of Southeast Alaska Karst: A Progress Report.' *Geoarchaeology* 12 (6):689–712. https://doi.org/10.1002/(SICI)1520-6548(199709)12:6<689::AID-GEA8>3.0.CO;2-V.

Drucker, Philip. 1955. 'Indians of the Northwest Coast.' In *Issue 10 of the Anthropological Handbook*. New York: McGraw-Hill, American Museum of Natural History.

Dunbar, James S., S. D. Webb, and Michael K. Faught. 1991. 'Inundated Prehistoric Sites in Apalachee Bay, Florida, and the Search for the Clovis Shoreline.' In *Paleoshorelines and Prehistory: An Investigation of Method*, edited by Lucille Lewis Johnson and Melanie Straight, 117–46. Ann Arbor: CRC Press.

Dunn, John A., and Arnold Booth. 1990. 'Tsimshian of Metlakatla, Alaska.' In *Handbook of North American Indians, Volume 7: Northwest Coast*, edited by Wayne

Suttles and William C. Sturtevant, 294–97. Washington, D.C.: Smithsonian Institution.

Dunnell, Robert C. 1992. 'The Notion Site.' In *Space, Time, and Archaeological Landscapes*, edited by Jacqueline Rossignol and LuAnn Wandsnider, 21–41. New York: Plenum Press.

Dunnell, Robert C., and William S Dancey. 1983. 'The Siteless Survey: A Regional Scale Data Collection Strategy.' In *Advances in Archaeological Method and Theory, Vol 6*, edited by Michael Brian Schiffer, 267–88. New York: Academic Press.

Easton, N. Alexander. 1993. *Underwater Archaeology in Montague Harbour: Interim Report on the 1992 Field Investigations*. Whitehorse: Occasional Papers of the Northern Research Institute.

Easton, N. Alexander, and C. Moore. 1991. 'Test Excavations of Subtidal Deposits at Montague Harbour, British Columbia.' *International Journal of Nautical Archaeology* 20:269–80.

Ebert, James I., and Timothy A. Kohler. 1988. 'The Theoretical Basis for Archaeological Predictive Modeling and a Consideration of Appropriate Data-Collection Methods.' In *Quantifying the Present and Predicting the Past: Theory, Method, and Application of Archaeological Predictive Modeling*, edited by W. James Judge and Lynne Sebastian, 97–171. Denver: US Department of the Interior, Bureau of Land Management.

Ejstrud, Bo. 2003. 'Indicative Models in Landscape Management: Testing the Methods.' *Symposium: The Archaeology of Landscape and Geographic Information Systems - Predictive Maps, Settlement Dynamics and Space and Territory in Prehistory*, 119–34.

Erlandson, Jon M. 2013. 'After Clovis-First Collapsed: Reimagining the Peopling of the Americas.' In *Paleoamerican Odyssey*, edited by Kelly E. Graf, Caroline V. Ketron, and Michael R. Waters, 127–32. Bryan: Center for the Study of the First Americans, Department of Anthropology, Texas A&M University.

Erlandson, Jon M., Michael H. Graham, Bruce J. Bourque, Debra Corbett, James A. Estes, and Robert S. Steneck. 2007. 'The Kelp Highway Hypothesis: Marine Ecology, the Coastal Migration Theory, and the Peopling of the Americas.' *Journal of Island and Coastal Archaeology* 2 (2):161–74. https://doi.org/10.1080/15564890701628612.

Erlandson, Jon M., Madonna L. Moss, and Matthew Des Lauriers. 2008. 'Life on the Edge: Early Maritime Cultures of the Pacific Coast of North America.' *Quaternary Science Reviews* 27 (23–24). Elsevier Ltd:2232–45. https://doi.org/10.1016/j.quascirev.2008.08.014.

Erlandson, Jon M., Todd J. Braje, Kristina M. Gill, and Michael H. Graham. 2015. 'Ecology of the Kelp Highway: Did Marine Resources Facilitate Human Dispersal From Northeast Asia to the Americas?' *Journal of Island and Coastal Archaeology* 10 (3):392–411. https://doi.org/10.1080/15564894.2014.1001923.

Erlandson, Jon M., Torben C. Rick, Todd J. Braje, Molly Casperson, Brendan Culleton, Brian Fulfrost, Tracy Garcia, et al. 2011. 'Paleoindian Seafaring, Maritime Technologies, and Coastal Foraging on California's Channel Islands.' *Science* 331 (March 4):1181–85. https://doi.org/10.1126/science.1196166.

ESRI. 2001. *ArcGIS 9: What Is ArcGIS?* USA: ESRI.

———. 2011. 'How Weighted Overlay Works.' How Weighted Overlay Works. 2011. http://help.arcgis.com/en/arcgisdesktop/10.0/help/index.html#//009z000000s1000000.htm.

———. 2012a. 'How High/Low Clustering (Getis-Ord General G) Works - Help |ArcGIS for Desktop.' ArcGIS Resource Center - ArcGIS Help Desktop 10er. 2012. http://help.arcgis.com/en/arcgisdesktop/10.0/help/index.html#//005p0000000q000000.

———. 2012b. 'How Spatial Autocorrelation (Global Moran's I) Works - Help |ArcGIS for Desktop.' ArcGIS Resource Center - ArcGIS Help Desktop 10. 2012. http://help.arcgis.com/en/arcgisdesktop/10.0/help/index.html#/How_Spatial_Autocorrelation_Global_Moran_s_I_works/005p0000000t000000/.

Faught, Michael K. 2004. 'Submerged Paleoindian and Archaic Sites of the Big Bend, Florida.' *Journal of Field Archaeology* 29 (3):273–90. https://doi.org/10.2307/3250893.

———. 2017. 'Where Was the PaleoAmerind Standstill?' *Quaternary International* 444. Elsevier Ltd:10–18. https://doi.org/10.1016/j.quaint.2017.04.038.

Faught, Michael K., and Amy E. Gusick. 2011. 'Submerged Prehistory in the Americas.' In *Submerged Prehistory*, edited by Jonathan Benjamin, Clive Bonsall, Catriona Pickard, and Anders Fischer, 145–57. Oxford: Oxbow Books. https://doi.org/10.13140/2.1.1029.6649.

Fedje, Daryl W. 2000. 'Millennial Tides and Archaeology in Gwaii Haanas.' *Research Links: A Forum for Natural, Cultural and Social Studies*, 2000.

———. 2003. 'Ancient Landscapes and Archaeology in Haida Gwaii, Hecate Strait.' *Archaeology of Coastal British Columbia: Essays in Honour of Professor Philip M Hobler*, 29–38.

Fedje, Daryl W., and Tina Christensen. 1999. 'Modeling Paleoshorelines and Locating Early Holocene Coastal Sites in Haida Gwaii.' *American Antiquity* 64 (4):635–52.

Fedje, Daryl W., Tina Christensen, Heiner Josenhans, Joanne B. Mcsporran, and Jennifer Strang. 2005. 'Millennial Tides and Shifting Shores: Archaeology on a Dynamic Landscape.' In *Haida Gwaii: Human History and Environment from the Time of the Loon to the Time of the Iron People*, edited by Daryl W. Fedje and Rolf W. Mathewes, 163–86. Vancouver: University of British Columbia Press.

Fedje, Daryl W., and Heiner Josenhans. 2000. 'Drowned Forests and Archaeology on the Continental Shelf of British Columbia, Canada.' *Geology* 28 (2):99–102. https://doi.org/10.1130/0091-7613(2000)28.

Fedje, Daryl W., Heiner Josenhans, John J. Clague, J. Vaughn Barrie, David J. Archer, and John R. Southon. 2005. 'Hecate Strait Paleoshorelines.' In *Haida Gwaii: Human History and Environment from the Time of the Loon to the Time of the Iron People*, edited by Daryl W. Fedje and Rolf W. Mathewes, 21–37. Vancouver: University of British Columbia Press.

Fedje, Daryl W., and Quentin Mackie. 2005. 'Overview of the Cultural History.' In *Haida Gwaii: Human History and Environment from the Time of the Loon to the Time of the Iron People*, edited by Daryl W. Fedje and Rolf W. Mathewes, 154–62. Vancouver: University of British Columbia Press.

Fedje, Daryl W., Alexander P. Mackie, Joanne B. Mcsporran, and B. Wilson. 1996. 'Early Period Archaeology in Gwaii Haanas: Results from the 1993 Field Program.' In *Early Human Occupation in British Columbia*, edited by Roy L. Carlson and Luke Della Bona, 133–50. Vancouver: University of British Columbia Press.

Fedje, Daryl W., Quentin Mackie, E. James Dixon, and Timothy H. Heaton. 2004. 'Late Wisconsin Environments and Archaeological Visibility on the Northern Northwest Coast.' In *Entering America: Northeast Asia and Beringia before the Last Glacial Maximum*, edited by David B. Madsen, 97–138. Salt Lake City: University of Utah Press.

Fedje, Daryl W., Alexander P. Mackie, Rebecca J. Wigen, Quentin Mackie, and Cynthia Lake. 2005. 'Kilgii Gwaay: An Early Maritime Site in the South of Haida Gwaii.' In *Haida Gwaii: Human History and Environment from the Time of the Loon to the Time of the Iron People*, edited by Daryl W. Fedje and Rolf W. Mathewes, 187–203. Vancouver: University of British Columbia Press.

Fedje, Daryl W., Quentin Mackie, Duncan McLaren, and Tina Christensen. 2008. 'Late Wisconsin Environments and Archaeological Visibility on the Northern Northwest Coast.' In *Projectile Point Sequences in Northwestern North America*, edited by Roy L. Carlson and Martin Magne, 19–40. Burnaby: Simon Fraser University Document Solutions.

Fedje, Daryl W., Martin Magne, and Tina Christensen. 2005. 'Test Excavations at Raised Beach Sites in Southern Haida Gwaii.' In *Haida Gwaii: Human History and Environment from the Time of the Loon to the Time of the Iron People*, edited by Daryl W. Fedje and Martin Magne, 204–44. Vancouver: University of British Columbia Press.

Fedje, Daryl W., and Rolf W. Mathewes. 2005. *Haida Gwaii: Human History and Environment from the Time of the Loon to the Time of the Iron People*. Edited by Daryl W. Fedje and Rolf W. Mathewes. Vancouver: University of British Columbia Press.

Fedje, Daryl W., Duncan McLaren, Thomas S. James, Quentin Mackie, Nicole F. Smith, John R. Southon, and Alexander P. Mackie. 2018. 'A Revised Sea Level History for the Northern Strait of Georgia, British Columbia, Canada.' *Quaternary Science Reviews* 192. The Authors:300–316. https://doi.org/10.1016/j.quascirev.2018.05.018.

Fedje, Daryl W., Joanne B. Mcsporran, and Andrew R. Mason. 1996. 'Early Holocene and Archaeology At the in Arrow Creek Site in Gwaiihaanas.' *Arctic Anthropology* 33 (1):116–42.

Fedje, Daryl W., Rebecca J. Wigen, Quentin Mackie, Cynthia Lake, and Ian D. Sumpter. 2001. 'Preliminary Results from Investigations at Kilgii Gwaay: An Early Holocene Archaeological Site on Ellen Island, Haida Gwaii, British Columbia.' *Canadian Journal of Archaeology* 25:98–120.

Feinman, Gary M. 1999. 'Defining a Contemporary Landscape Approach: Concluding Thoughts.' *Antiquity* 73 (281):684–85. https://doi.org/10.1111/j.1558-5646.2011.01242.x/abstract.

Fischer, Anders. 1995a. 'An Entrance to the Mesolithic World below the Ocean. Status of Ten Years' Work on the Danish Sea Floor.' Edited by Anders Fischer. *Man and Sea in the Mesolithic*, no. April. Oxford: Oxbow Monograph 53:371–84.

———. 1995b. 'Epilogue to the Man & Sea Symposium.' In *Man and Sea in the Mesolithic: Coastal Settlement above and below Present Sea Level*, edited by Anders Fischer, 431–35. Oxford: Oxbow Monograph 53. https://doi.org/10.1063/1.460797.

———. 2004. 'Submerged Stone Age - Danish Examples and North Sea Potential.' In *Submarine Prehistoric Archaeology of the North Sea: Research Priorities and Collaboration with Industry*, edited by Nicholas C. Flemming, 21–36. York: CBA Research Report 141, English Heritage, Council for British Archaeology.

———. 2011. 'Stone Age on the Continental Shelf: An Eroding Resource.' In *Submerged Prehistory*, edited by Jonathan Benjamin, Clive Bonsall, Catriona Pickard, and Anders Fischer, 298–310. Oxford: Oxbow Books.

Fischer, Christopher R., and Tina L. Thurston. 1999. 'Dynamic Landscapes and Socio-Political Process: The Topography of Anthropogenic Environments in Global Perspective.' *Antiquity* 73 (281):630. https://doi.org/10.1111/j.1558-5646.2011.01242.x/abstract.

Fischer, George. 1974. 'The History and Nature of Underwater Archaeology in the National Park Service.' In *Underwater Archaeology in the National Park Service: A Model for the Management of Submerged*

Cultural Resources, edited by Daniel J. Leniham, 3–8. Washington, D.C.: Department of the Interior, US National Park Service.

Fitzhugh, William. 1975. 'A Comparative Approach to Northern Maritime Adaptations.' In *Prehistoric Maritime Adaptations of the Circumpolar Zone*, edited by William Fitzhugh, 339–86. Paris: Mouton Publishers.

Fladmark, Knut R. 1975. *A Paleoecological Model of Northwest Coast Prehistory*. Ottawa: National Museum of Man, Mercury Series, Archaeological Survey of Canada Paper No 43.

———. 1979. 'Alternate Migration Corridors for Early Man in North America.' *American Antiquity* 44 (1):55–69.

———. 1982. 'An Introduction to the Prehistory of British Columbia.' *Canadian Journal of Archaeology* 6:95–155. http://www.env.gov.bc.ca/wld/documents/techpub/rn324.pdf.

Flannery, Kent V. 1968. 'Archaeological System Theory and Early Mesoamerica.' In *Anthropological Archaeology in the Americas*, edited by Betty J. Meggers, 67–87. Washington, D.C.: Anthropological Society of Washington.

Flatman, Joseph C. 2003. 'Cultural Biographies, Cognitive Landscapes and Dirty Old Bits of Boat: 'Theory' in Maritime Archaeology.' *International Journal of Nautical Archaeology* 32 (2):143–57. https://doi.org/10.1016/j.ijna.2003.07.001.

———. 2011. 'Places of Special Meaning: Westerdahl's Comment, 'Agency', and the Concept of the 'Maritime Cultural Landscape.' In *The Archaeology of Maritime Landscapes*, edited by Ben Ford, 311–30. New York: Springer.

Flatman, Joseph C., and Amanda Evans. 2014. 'Prehistoric Archaeology on the Continental Shelf: The State of the Science in 2013.' In *Prehistoric Archaeology on the Continental Shelf: A Global Review*, edited by Amanda Evans, Joseph C. Flatman, and Nicholas C. Flemming, 1–12. New York: Springer.

Fleming, Andrew. 2006. 'Post-Processual Landscape Archaeology: A Critique.' *Cambridge Archaeology Journal* 16 (3):267–80.

Fleming, M.A., and J.A. Cook. 2002. 'Phylogeography of Endemic Ermine (*Mustela Erminea*) in Southeast Alaska.' *Molecular Ecology* 11:795–807.

Flemming, Nicholas C. 2017. 'The Role of the Submerged Prehistoric Landscape in Ground-Truthing Models of Human Dispersal during the Last Half Million Years.' In *Under the Sea: Archaeology and Palaeolandscapes of the Continental Shelf*, edited by Geoffrey N. Bailey, Jan Harff, and Dimitris\ Sakellariou, 269–83. Cham, Switzerland: Springer, Coastal Research Library, Volume 20. https://doi.org/10.1007/978-3-319-53160-1.

Ford, Ben. 2011a. 'Introduction.' In *The Archaeology of Maritime Landscapes*, edited by Ben Ford, 1–9. New York: Springer.

———. 2011b. 'The Shoreline as a Bridge, Not a Boundary: Cognitive Maritime Landscapes of Lake Ontario.' In *The Archaeology of Maritime Landscapes*, edited by Ben Ford, 63–80. New York: Springer.

Ford, Ben, Jessi J. Halligan, and Jonathan Benjamin. 2010. 'A North American Perspective: Comment on Jonathan Benjamin's 'Submerged Prehistoric Landscapes and Underwater Site Discovery: Reevaluating the "Danish Model" for International Practice.' *Journal of Island and Coastal Archaeology* 5 (2):277–79. https://doi.org/10.1080/15564894.2010.506623.

Forrest, C. 2002. 'Defining 'Underwater Cultural Heritage.' *The International Journal of Nautical Archaeology* 31 (1):3–11. https://doi.org/10.1006/ijna.2002.1022.

Freeman, Andrea. 2016. 'Why the Ice-Free Corridor Is Still Relevant to the Peopling of the New World.' In *Stones, Bones and Profiles: Exploring Archaeological Context*, edited by M Kornfeld and Bruce B. Huckell, 51–74. Boulder: University Press of Colorado.

Gaffney, Vincent, and Kenneth Thomson. 2007. 'Mapping Doggerland.' In *Mapping Doggerland: The Mesolithic Landscapes of the Southern North Sea*, edited by Vincent Gaffney, Kenneth Thomson, and Simon Fitch, 1–10. Oxford: Archaeopress.

Gaffney, Vincent, Robin Allaby, Richard Bates, Martin Bates, Eugene Ch, Simon Fitch, Paul Garwood, et al. 2017. 'Doggerland and the Lost Frontiers Project (2015-2010).' In *Under the Sea: Archaeology and Palaeolandscapes of the Continental Shelf*, edited by Geoffrey N. Bailey, Jan Harff, and Dimitris Sakellariou, 305–19. Cham, Switzerland: Springer, Coastal Research Library, Volume 20. https://doi.org/10.1007/978-3-319-53160-1.

Garrison, Ervan G., Jessica Cook Hale, Christopher Sean Cameron, and Erin Smith. 2016. 'The Archeology, Sedimentology and Paleontology of Gray's Reef National Marine Sanctuary and Nearby Hard Bottom Reefs along the Mid Continental Shelf of the Georgia Bight.' *Journal of Archaeological Science: Reports* 5. Elsevier Ltd:240–62. https://doi.org/10.1016/j.jasrep.2015.11.009.

Gehrels, G. E., and H. C. Berg. 1992. 'Geologic Map of Southeast Alaska.' *U.S. Geological Survey Miscellaneous Investigations MSeries Map I-1867*.

Getis, Arthur, and J. K. Ord. 1992. 'The Analysis of Spatial Association by Use of Distance Statistics.' *Geographical Analysis* 24 (3):189–206.

Gibson, James R. 1992. *Otter Skins, Boston Ships, and China Goods: The Maritime Fur Trade of the Northwest Coast, 1785-1841*. Seattle: University of Washington.

Gilbert, M. Thomas P., Dennis L. Jenkins, Anders Götherstrom, Nuria Naveran, Juan J. Sanchez, Michael Hofreiter, Philip Francis Thomsen, et al. 2008. 'DNA from Pre-Clovis Human Coprolites in Oregon, North America.' *Science* 320 (5877):786–89. https://doi.org/10.1126/science.1154116.

Gljjrstad, Hakon, Jostein Gundersen, and K Frode. 2017. 'The Northern Coasts of Doggerland and the Colonisation of Norway at the End of the Ice Age.' In *Under the Sea: Archaeology and Palaeolandscapes of the Continental Shelf*, edited by Geoffrey N. Bailey, Jan Harff, and Dimitris Sakellariou, Coastal Re, 285–303. Cham, Switzerland: Springer. https://doi.org/10.1007/978-3-319-53160-1_19.

Glørstad, Håkon. 2014. 'Deglaciation, Sea-Level Change and the Holocene Colonization of Norway.' *Geological Society, London, Special Publications* 411:2005–7. https://doi.org/10.1144/SP411.7.

Goebel, T., and S. B. Slobodin. 1999. 'The Colonization of Western Beringia: Technology, Ecology, and Adaptations.' *Ice Age People of North America: Environments, Origins, and Adaptations*, 104–55.

Goggin, John M. 1960. 'Underwater Archaeology: Its Nature and Limitations.' *American Antiquity* 25 (3):348–54.

Gormley, Mary. 1971. 'Tlingits of Bucareli Bay, Alaska (1774-1792).' *Northwest Anthropological Research Notes* 5 (2):157–80.

Gould, Richard A. 2000. *Archaeology and the Social History of the Ship*. Cambridge: Cambridge University Press.

GRASS Development Team. 2015. 'Geographic Resources Analysis Support System (GRASS) Software, Version 7.' Open Source Geospatial Foundation. 2015. https://grass.osgeo.org.

Green, Ernestene L. 1973. 'Location Analysis of Prehistoric Maya Sites in Northern British Honduras.' *American Antiquity* 38 (3):279–93.

Green, Jeremy. 2003. *Maritime Archaeology: A Technical Handbook*. London: Academic Press.

Green, Jeremy, and Matthew Gainsford. 2003. 'Evaluation of Underwater Surveying Techniques.' *International Journal of Nautical Archaeology* 32 (2):252–61. https://doi.org/10.1016/j.ijna.2003.08.007.

Gruhn, Ruth. 1994. 'The Pacific Coast Route of Initial Entry: An Overview.' In *Method and Theory for Investigating the Peopling of the Americas. Clovis Origins and Adaptations*, edited by Robson. Bonnichsen and D. Gentry. Steele, 249–56. Corvallis: Center for the Study of the First Americans, Oregon State University.

Guisck, Amy E., and Michael K. Faught. 2011. 'Prehistoric Archaeology Underwater: A Nascent Subdiscipline Critical to Understanding Early Coastal Occupations and Migration Routes.' In *Trekking the Shore*, edited by Nuno F. Bicho, Jonathan Haws, and Loren G. Davis, 27–50. Dordrecht: Springer.

Halligan, Jessi J. 2011. 'Lake Ontario Paleoshorelines and Submerged Prehistoric Site Potential in Great Lakes.' In *The Archaeology of Maritime Landscapes*, edited by Ben Ford, 45–62. New York: Springer.

Halligan, Jessi J., Michael R. Waters, Angelina Perrotti, Ivy J. Owens, Joshua M. Feinberg, Mark D. Bourne, Brendan Fenerty, et al. 2016. 'Pre-Clovis Occupation 14,550 Years Ago at the Page-Ladson Site, Florida, and the Peopling of the Americas.' *Science Advances* 2 (5):e1600375. https://doi.org/10.1126/sciadv.1600375.

Halpin, Marjorie M., and Margaret Seguin. 1990. 'Tsimshian Peoples: Southern Tsimshian, Coast Tsimshian, Nishga, and Gitkan.' In *Handbook of North American Indians, Volume 7: Northwest Coast*, edited by Wayne Suttles and William C. Sturtevant, 267–84. Washington, D.C.

Hamilton, Scott, and Linda Larcombe. 1994. 'Cultural Heritage Resource Predictive Modelling Project: Volume 1: Introduction to the Research.' Thunder Bay, Ontario.

Hansen, Barbara C. S., and Daniel R. Engstrom. 1996. 'Vegetation History of Pleasant Island, Southeastern Alaska, since 13,000 Yr B.P.' *Quaternary Research* 46 (2):161–75. https://doi.org/10.1006/qres.1996.0056.

He, F. 2011. 'Simulating Transient Climate Evolution of the Last Deglaciation with CCSM3.' University of Wisconsin, Madison.

Heaton, Timothy H., and Frederick Grady. 2003. 'The Late Wisconsin Vertebrate History of Prince of Wales Island, Southeast Alaska.' In *Ice Age Cave Faunas of North America*, edited by J L Mead and R W Graham, 17–53. Bloomington: Indiana University Press.

Heusser, Calvin J. 1955. 'Pollen Profiles from the Queen Charlotte Islands, British Columbia.' *Canadian Journal of Botany* 33 (5):429–49. https://doi.org/https://doi-org.ezproxy.lib.ucalgary.ca/10.1139/b55-036.

———. 1960. *Late-Pleistocene Environments of North Pacific North America: An Elaboration of Late-Glacial and Postglacial Climatic, Physiographic, and Biotic Changes*. New York: American Geographic Society.

Hobler, Philip M. 1978. 'The Relationship of Archaeological Sites to Sea Levels on Moresby Island, Queen Charlotte Islands.' *Canadian Journal of Archaeology* 2 (2):1–14.

Hodder, Ian. 1977. 'Some New Directions in the Spatial Analysis of Archaeological Data at the Regional Scale (Macro).' In *Spatial Archaeology*, edited by David L. Clarke, 223–330. New York: Academic Press, Inc.

Hoffecker, John F., Scott A. Elias, and Dennis H. O. Rourke. 2014. 'Out of Beringia?' *Science Perspectives* 343 (February):979–80.

Hoffecker, John F., Scott A. Elias, Dennis H. O'Rourke, G. Richard Scott, and Nancy H. Bigelow. 2016. 'Beringia and the Global Dispersal of Modern Humans.' *Evolutionary Anthropology* 25 (2). https://doi.org/10.1002/evan.21478.

Imagenex. 2011. 'Imagenex Model 872 'Yellowfin' Sidescan Sonar.' Imagenex Technology Corp. 2011. http://www.imagenex.com/YellowFin_manual_v2.pdf.

Ingold, Tim. 1993. 'The Temporality of the Landscape.' *World Archaeology* 25 (2):152–74.

Intellicast. 2018. 'Ketchikan - Historic Average Weather.' Intellicast: The Authority in Expert Weather. 2018. http://www.intellicast.com/Local/History.aspx?location=USAK0125.

Ives, John W., Duane G. Froese, Kisha Supernant, and Gabriel Yanicki. 2013. 'Vectors, Vestiges, and Vallhallas - Rethinking the Corridor.' In *Paleoamerican Odyssey*, edited by Kelly E. Graf, Caroline V. Ketron, and Michael R. Waters, 149–70. Bryan: Center for the Study of the First Americans, Department of Anthropology, Texas A&M University.

Johnson, G. A. 1977. 'Aspects of Regional Analysis in Archaeology.' *Annual Review of Anthropology* 6 (1977):479–508. https://doi.org/10.1146/annurev.an.06.100177.002403.

Johnson, James A., Brad A. Andres, and John A. Bissonette. 2008. 'Birds of the Major Mainland Rivers of Southeast Alaska.' *U.S. Forest Service General Technical Report PNW* 739 (July):1–88.

Josenhans, Heiner, Daryl W. Fedje, R. Pienitz, and John R. Southon. 1997. 'Early Humans and Rapidly Changing Holocene Sea Levels in the Queen Charlotte Islands Hecate Strait, British Columbia, Canada.' *Science* 277 (5322):71–74. https://doi.org/10.1126/science.277.5322.71.

Kantner, John. 2008. 'The Archaeology of Regions: From Discrete Analytical Toolkit to Ubiquitous Spatial Perspective.' *Journal of Archaeological Research* 16 (1):37–81. https://doi.org/10.1007/s10814-007-9017-8.

Kari, James, and Ben A. Potter. 2010. 'The Dene-Yeniseian Connection: Bridging Asia and North America.' In *The Dene-Yeneiseian Connection: Bridging Asia and North America*, edited by James Kari and Ben A. Potter, 1–24. Fairbanks: Department of Anthropology, University of Alaska, New Series, Vol 5 (1-2).

Kemp, Brian M., Ripan S. Malhi, John McDonough, Deborah A. Bolnick, Jason A. Eshleman, Olga Rickard, Cristina Martinez-Labarga, et al. 2007. 'Genetic Analysis of Early Holocene Skeletal Remains from Alaska and Its Implications for the Settlement of the Americas.' *American Journal of Physical Anthropology* 132:605–21. https://doi.org/10.1002/ajpa.

Kincaid, C. 1988. 'Predictive Modeling and Its Relationship to Cultural Resource Management.' In *Quantifying the Present and Predicting the Past: Theory, Method, and Application of Archaeological Predictive Modeling*, edited by W. James Judge and Lynne Sebastian, 549–80. Denver: US Department of the Interior, Bureau of Land Management.

King, A. 1950. *Cattle Point, A Stratified Site on the Southern Northwest Coast*. Salt Lake City: Memoirs of the Society for American Archaeology, 7.

Knapp, A. Bernard, and Wendy Ashmore. 1999. 'Archaeological Landscapes: Constructed, Conceptualized, Ideational.' In *Archaeologies of Landscapes: Contemporary Perspectives*, edited by Wendy Ashmore and A. Bernard Knapp, 1–30. Oxford: Blackwell Publishers.

Kohler, Timothy A. 1988. 'Predictive Locational Modeling: History and Current Practice.' In *Quantifying the Present and Predicting the Past: Theory, Method, and Application of Archaeological Predictive Modeling*, edited by W. James Judge and Lynne Sebastian, 19–96. Denver: US Department of the Interior, Bureau of Land Management.

Kohler, Timothy A., and Sandra C. Parker. 1986. 'Predictive Models for Archaeological Resource Location.' In *Advances in Archaeological Method and Theory, Vol 9*, edited by Michael Brian Schiffer, 397–452. New York: Academic Press, Inc.

Kondo, Yasuhisa, and Takashi Oguchi. 2012. 'How Can We Apply Ecological Niche Models to Palaeoanthropological Research?' Tokyo.

Kondo, Yasuhisa, Takayuki Omori, and Philip Verhagen. 2012. 'Developing Predictive Models for Palaeoanthropological Research: A Preliminary Discussion.' *Technical Report, Department of Computer Science, Tokyo Institute of Technology*, 10. https://doi.org/urn:NBN:nl:ui:31-1871/43686.

Kongsberg. 2006. 'EM 3002 Multibeam Echo Sounder: The New Generation High Performance Shallow Water Multibeam.' Horten: Kongsberg Marine, 3000–3003. https://www.km.kongsberg.com/ks/web/nokbg0397.nsf/AllWeb/7C8510CFA3CD21ABC1256CF00052DD1C/$file/164771ae_EM3002_Product_spec_lr.pdf.

Kroeber, Alfred L. 1923. 'American Culture and the Northwest Coast.' *American Anthropologist* 25 (1):1–20.

———. 1939. *Cultural and Natural Areas of North America*. Berkeley: University of California Publications in American Archaeology and Ethnology 38.

Kvamme, Kenneth L. 1988a. 'Development and Testing of Quantitative Models.' In *Quantifying the Present and*

Predicting the Past: Theory, Method, and Application of Archaeological Predictive Modeling, edited by W. James Judge and Lynne Sebastian, 325–428. Denver: US Department of the Interior, Bureau of Land Management.

———. 1988b. 'Using Existing Data for Model Building.' In *Quantifying the Present and Predicting the Past: Theory, Method, and Application of Archaeological Predictive Modeling*, edited by W. James Judge and Lynne Sebastian, 301–23. Denver: US Department of the Interior, Bureau of Land Management.

———. 2006. 'There and Back Again: Revisiting Archaeological Locational Modeling.' In *GIS and Archaeological Site Location Modeling*, edited by Mark W. Mehrer and Konnie L. Wescott, 2–34. New York: Taylor and Francis.

Lacourse, Terri, and Rolf W. Mathewes. 2005. 'Terrestrial Paleoecology of Haida Gwaii and the Continental Shelf: Vegetation, Climate, and Plant Resources of the Coastal Migration Route.' In *Haida Gwaii: Human History and Environment from the Time of the Loon to the Time of the Iron People*, edited by Daryl W. Fedje and Rolf W. Mathewes, 38–58. Vancouver: University of British Columbia Press.

Lacourse, Terri, Rolf W. Mathewes, and Daryl W. Fedje. 2005. 'Late-Glacial Vegetation Dynamics of the Queen Charlotte Islands and Adjacent Continental Shelf, British Columbia, Canada.' *Palaeogeography, Palaeoclimatology, Palaeoecology* 226 (1–2):36–57. https://doi.org/10.1016/j.palaeo.2005.05.003.

Lacroix, Dominic, Trevor Bell, John Shaw, and Kieran Westley. 2014. 'Submerged Archaeological Landscapes and the Recording of Precontact History: Examples from Atlantic Canada.' In *Prehistoric Archaeology on the Continental Shelf: A Global Review*, edited by Amanda Evans, Joseph C. Flatman, and Nicholas C. Flemming, 13–36. New York: Springer.

Laguna, Frederica de. 1960. *The Story of a Tlingit Community: A Problem Inn the Relationship between Archaeological, Ethnological, and Historical Methods*. Washington, D.C.: Bureau of American Ethnology, Bulletin 172.

———. 1990. 'Tlingit.' In *Handbook of North American Indians, Volume 7: Northwest Coast*, edited by Wayne Suttles and William C. Sturtevant, 203–28. Washington, D.C.: Smithsonian Institution Press.

Langdon, Stephen John. 1977. 'Technology, Ecology and Economy: Fishing Systems in Southeast Alaska.' Stanford University. https://doi.org/10.16953/deusbed.74839.

Lee, Craig Michael. 2007. 'Origin and Function of Early Holocene Microblade Technology in Southeast Alaska, USA.' University of Colorado, Boulder.

Lesnek, Alia J., Jason P. Briner, Charlotte Lindqvist, James F. Baichtal, and Timothy H. Heaton. 2018. 'Deglaciation of the Pacific Coastal Corridor Directly Preceded the Human Colonization of the Americas.' *Science Advances* 4 (5):eaar5040. https://doi.org/10.1126/sciadv.aar5040.

Letham, Bryn, Andrew Martindale, Duncan McLaren, Brown Thomas, Kenneth M. Ames, David Archer, and Susan Marsden. 2015. 'Holocene Settlement History of the Dundas Islands Archipelago, Northern British Columbia.' *BC Studies* 187 (Autumn):51-60,626-65,69-85,303-305. https://doi.org/http://dx.doi.org/10.1108/17506200710779521.

Letham, Bryn, Andrew Martindale, Rebecca Macdonald, Eric Guiry, Jacob Jones, and Kenneth M. Ames. 2016. 'Postglacial Relative Sea-Level History of the Prince Rupert Area, British Columbia, Canada.' *Quaternary Science Reviews* 153. Elsevier Ltd:156–91. https://doi.org/10.1016/j.quascirev.2016.10.004.

Letham, Bryn, Andrew Martindale, Nicholas Waber, and Kenneth M. Ames. 2018. 'Archaeological Survey of Dynamic Coastal Landscapes and Paleoshorelines: Locating Early Holocene Sites in the Prince Rupert Harbour Area, British Columbia, Canada.' *Journal of Field Archaeology* 43 (3). Taylor & Francis:181–99. https://doi.org/10.1080/00934690.2018.1441575.

Leusen, Martijn P. van, Jos Deeben, Daan Hallewas, Paul Zoetbrood, Hans Kamermans, and Philip Verhagen. 2005. 'A Baseline for Predictive Modelling in the Netherlands.' In *Predictive Modelling for Archaeological Heritage Management: A Research Agenda*, edited by Hans Kamermans and Martijn van Leusen, 25–92. Amersfoort: Rijksdienst voor het Oudheidkundig Bodemonderzoek.

Lewis, David. 1972. *We, the Navigators: The Asncient Art of Landfinding in the Pacific*. Honolulu: University of Hawaii Press.

Lindo, John, Alessandro Achilli, Ugo A. Perego, David Archer, Cristina Valdiosera, Barbara Petzelt, Joycelynn Mitchell, et al. 2017. 'Ancient Individuals from the North American Northwest Coast Reveal 10,000 Years of Regional Genetic Continuity.' *Proceedings of the National Academy of Sciences* 114 (16):4093–98. https://doi.org/10.1073/pnas.1620410114.

Liu, Z., L. Otto-Bliesner, F. He, E. C. Brady, R. Tomas, P. U. Clark, A. E. Carlson, et al. 2009. 'Transient Simulation of Last Deglaciation with a New Mechanism for Bolling-Allerod Warming.' *Science* 325 (5938):310–14.

Lock, Gary. 2004. 'Wetland Archaeology and Computers.' In *Living on the Lake in Prehistoric Europe: 150 Years of Lake-Dwelling Research*, edited by Francesco Menotti, 194–204. New York: Routledge.

Lock, Gary, and Trevor Harris. 1992. 'Visualizing Spatial Data: The Importance of Geographic Information Systems.' In *Archaeology and the Information Age: A*

Global Perspective, edited by Paul Reilly and Sebastian Rahtz, 81–96. London: Routledge.

———. 2006. 'Enhancing Predictive Archaeological Modeling: Integrating Location, Landscape, and Culture.' In *GIS and Archaeological Site Location Modeling*, edited by Mark W. Mehrer and Konnie L. Wescott, 36–55. New York: Taylor and Francis.

Longley, Paul A., Michael F. Goodchild, David J. Maguire, and David W. Rhind. 2005. *Geographic Information Systems and Science, Second Edition*. Chichester: John Wiley & Sons, Ltd.

Lund, Carsten. 1995. 'Legislation, Protection and Management of Underwater Cultural Heritage - the Danish Model.' In *Man and Sea in the Mesolithic: Coastal Settlement above and below Present Sea Level*, edited by Anders Fischer, 415–18. Oxford: Oxbow Monograph 53.

Lyman, R. Lee. 1991. *Prehistory of the Oregon Coast: The Effects of Excavation Strategies and Assemblage Size on Archaeological Inquiry*. San Diego: Academic Press.

Lyons, C. P., and Bill Merilees. 1995. *Trees, Shrubs, and Flowers to Know in Washington and British Columbia*. Vancouver: Long Pine Publishing.

MacDonald, G. M., D. W. Beilman, Y. V. Kuzmin, L. A. Orlova, K. V. Kremenetski, B. Shapiro, R. K. Wayne, and B. Van Valkenburgh. 2012. 'Pattern of Extinction of the Woolly Mammoth in Beringia.' *Nature Communications* 3 (May). Nature Publishing Group:893–98. https://doi.org/10.1038/ncomms1881.

MacDonald, Stephen O., and Joseph A. Cook. 2009. *Recent Mammals of Alaska*. Fairbanks: University of Alaska Press.

MacKenzie, J. D. 1916. 'Geology of Graham Island, British Columbia.' *Geological Survey of Canada* Memoir 88.

Mackie, Quentin, Daryl W. Fedje, and Duncan McLaren. 2018. 'Archaeology and Sea Level Change on the British Columbia Coast.' *Canadian Journal of Archaeology* 42 (July):74–91.

Mackie, Quentin, Daryl W. Fedje, Duncan McLaren, Nicole Smith, and Iain McKechnie. 2011. 'Early Environments and Archaeology of Coastal British Columbia.' In *Trekking the Shore: Changing Coastlines and the Antiquity of Coastal Settlement*, edited by Nuno F. Bicho, Jonathan A. Haws, and Loren G. Davis, 51–103. Dordrecht: Interdisciplinary contributions to archaeology, Springer. https://doi.org/10.1007/978-1-4419-8219-3.

Mackie, Alexander P., and Ian D. Sumpter. 2005. 'Shoreline Settlement Patterns in Gwaii Haanas during the Early and Late Holocene.' In *Haida Gwaii: Human History and Environment from the Time of the Loon to the Time of the Iron People*, edited by Daryl W. Fedje and Rolf W. Mathewes, 337–71. Vancouver: University of British Columbia Press.

Mandryk, Carole A. S., Heiner Josenhans, Daryl W. Fedje, and Rolf W. Mathewes. 2001. 'Late Quaternary Paleoenvironments of Northwestern North America: Implications for Inland versus Coastal Migration Routes.' *Quaternary Science Reviews* 20 (1–3):301–14. https://doi.org/10.1016/S0277-3791(00)00115-3.

Mann, Daniel H. 1986. 'Wisconsin and Holocene Glaciation of Southeast Alaska.' In *Glaciation in Alaska - The Geologic Record*, edited by Thomas D. Hamilton, Katherine M. Reed, and Robert M. Thorson, 237–65. Anchorage: Alaska Geological Society.

Mann, Daniel H., and Thomas D. Hamilton. 1995. 'Late Pleistocene and Holocene Paleoenvironments of the North Pacific Coast.' *Quaternary Science Reviews* 14:449–71.

Mann, Daniel H., and Gregory P. Streveler. 2008. 'Post-Glacial Relative Sea Level, Isostasy, and Glacial History in Icy Strait, Southeast Alaska, USA.' *Quaternary Research* 69 (2):201–16. https://doi.org/10.1016/j.yqres.2007.12.005.

Martindale, Andrew, and Kisha Supernant. 2009. 'Quantifying the Defensiveness of Defended Sites on the Northwest Coast of North America.' *Journal of Anthropological Archaeology* 28 (2). Elsevier Inc.:191–204. https://doi.org/10.1016/j.jaa.2009.01.001.

Maschner, Herbert D. G. 1992. 'The Origins of Hunter and Gatherer Sedentism and Political Complexity: A Case Study from the Northern Northwest Coast.' University of California Santa Barbara.

———. 1996. 'Geographic Information Systems in Archaeology.' In *New Methods, Old Problems: Geographic Information Systems in Modern Archaeological Research*, edited by Herbert D. G. Maschner, 1–24. Carbondale: Occasional Paper No 23, Center for Archaeological Investigations, Southern Illinois University.

Maschner, Herbert D. G., and Jeffrey W. Stein. 1995. 'Multivariate Approaches to Site Location on the Norwest Coast of North America.' *Antiquity* 69:61–73.

Masselink, Gerhard, and Michael G. Hughes. 2003. *Introduction to Coastal Processes and Geomorphology*. New York: Oxford University Press Inc.

Masters, Patricia. 1983. 'Detection and Assessment of Prehistoric Artifact Sites Off the Coast of Southern California.' In *Quaternary Coastline and Marine Archaeology: Towards the Prehistory of Land Bridges and Continental Shelves*, edited by Patricia M. Masters and Nicholas C. Flemming, 189–214. New York: Academic Press.

Mathews, William. 1979. 'Late Quaternary Environmental History Affecting Human History of the Pacific

Northwest.' *Canadian Journal of Archaeology* 3 (3):145–56. J10.3.

Matson, R. G., and Gary Coupland. 1995. *The Prehistory of the Northwest Coast*. San Diego: Academic Press.

Mazurkevich, Andrey, and Ekaterina Dolbounova. 2011. 'Underwater Investigations in Northwest Russia: Lacustrine Archaeology of Neolithic Pile Dwellings.' In *Submerged Prehistory*, edited by Johnathan Benjamin, Clive Bonsall, Catriona Pichard, and Anders Fischer, 158–72. Oxford: Oxbow Books.

McCartney, Allen P. 1974. 'Maritime Adaptations on the North Pacific Rim.' *Arctic Anthropology* 11 (1974):153–62.

McCoy, Mark D., and Thegn N. Ladefoged. 2009. 'New Developments in the Use of Spatial Technology in Archaeology.' *Journal of Archaeological Research* 17 (3):263–95. https://doi.org/10.1007/s10814-009-9030-1.

McGrail, Sean. 1981. *The Ship: Rafts, Boats and Ships*. London: Her Majesty's Stationery Office.

McLaren, Duncan, Quentin Mackie, Daryl W. Fedje, Andrew Martindale, and David Archer. 2007. 'Relict Shorelines and Archaeological Prospection on the Continental Hinge of North Coastal British Columbia.' *72nd Annual Meeting of the Society for American Archaeology*, 1–23.

McLaren, Duncan, Daryl W. Fedje, Murray B. Hay, Quentin Mackie, Ian J. Walker, Dan H. Shugar, Jordan B.R. Eamer, Olav B. Lian, and Christina Neudorf. 2014. 'A Post-Glacial Sea Level Hinge on the Central Pacific Coast of Canada.' *Quaternary Science Reviews* 97. Elsevier Ltd:148–69. https://doi.org/10.1016/j.quascirev.2014.05.023.

McLaren, Duncan, Farid Rahemtulla, Elroy White Gitlas, and Daryl W. Fedje. 2015. 'Prerogatives, Sea Level, and the Strength of Persistent Places: Archaeological Evidence for Long-Term Occupation of the Central Coast of British Columbia.' *BC Studies*, no. 187:155–67, 174–81, 183–91, 303, 305–6. https://doi.org/http://dx.doi.org/10.1108/17506200710779521.

McLaren, Duncan, Daryl W. Fedje, Angela Dyck, Quentin Mackie, Alisha Gauvreau, and Jenny Cohen. 2018. 'Terminal Pleistocene Epoch Human Footprints from the Pacific Coast of Canada.' Edited by Michael D. Petraglia. *PLOS ONE* 13 (3):e0193522. https://doi.org/10.1371/journal.pone.0193522.

McNiven, Ian J. 2008. 'Sentient Sea : Seascapes as Spiritscapes.' In *Handbook of Landscape Archaeology*, edited by Bruno David and Julian Thomas, 149–57. Walnut Creek, CA: Left Coast Press.

Meares, John. 1790. *Voyages Made in the Years 1788, and 1789 from China to the North West Coast of America*. London: Logographic press.

Mehrer, Mark W. 2006. 'Preface.' In *GIS and Archaeological Site Location Modeling*, edited by Mark W. Mehrer and Konnie L. Wescott, ix–x. New York: Taylor and Francis.

Meighan, Clement W. 1983. 'Early Man in the New World.' In *Quaternary Coastline and Marine Archaeology: Towards the Prehistory of Land Bridges and Continental Shelves*, edited by Patricia M. Masters and Nicholas C. Flemming, 441–62. London: Academic Press.

MerriamWebster. 2018. 'Merriam Webster Dictionary Online.' Merriam Webster Dictionary Online. 2018. https://www.merriam-webster.com.

Metz, M, H. Mitasova, and R. S. Harmon. 2011. 'Efficient Extraction of Drainage Networks from Passive, Radar-Based Elevation Model with Least Cost Path Search.' *Hydrology and Earth System Sciences* 15:667–78.

Mink II, Philip B., B. Jo Stokes, and David Pollack. 2006. 'Points vs. Polygons: A Test Case Using a Statewide Geographic Information System.' In *GIS and Archaeological Site Location Modeling*, edited by Mark W. Mehrer and Konnie L. Wescott, 200–219. New York: Taylor and Francis.

Momber, Garry. 2000. 'Drowned and Deserted: A Submerged Prehistoric Landscape in the Solent, England.' *International Journal of Nautical Archaeology* 29 (1):86–99. https://doi.org/10.1006/ijna.2000.0284.

———. 2011. 'Submerged Landscape Excavations in the Solent, Southern Britain: Climate Change and Cultural Development.' In *Submerged Prehistory*, edited by Jonathan Benjamin, Clive Bonsall, Catriona Pickard, and Anders Fischer, 85–98. Oxford: Oxbow Books.

Monteleone, Kelly Rose Bale. 2013. 'Lost Worlds: Locating Submerged Archaeological Sites in Southeast Alaska.' University of New Mexico.

———. 2016. 'Exploring the Consistency of Archaeological Site Locations in the Prince of Wales Area in Southeast Alaska Over the Last 5000 Years.' *Alaska Journal of Anthropology* 14 (1&2):1–15.

Monteleone, Kelly Rose Bale, and E. James Dixon. 2018. 'Gateway to the Americas II: Permit Final Report 2012:01.' Anchorage.

Monteleone, Kelly Rose Bale, and Andrew D. Wickert. 2013. 'Investigating the Potential for Archaeological Sites on the Submerged Southern Beringian Archipelago.' In *Poster Presented at Paleoamerican Odyssey*. Santa Fe.

Monteleone, Kelly Rose Bale, Andrew D. Wickert, and E. James Dixon. 2017. 'Underwater Archaeological Surveys in Shakan Bay, SE Alaska.' Vancouver.

Moreno-Mayar, J. Víctor, Ben A. Potter, Lasse Vinner, Matthias Steinrücken, Simon Rasmussen, Jonathan Terhorst, John A. Kamm, et al. 2018. *Terminal Pleistocene Alaskan Genome Reveals First Founding*

Population of Native Americans. Nature. Vol. 553. https://doi.org/10.1038/nature25173.

Moss, Madonna L. 1992. 'Relationships between Maritime Cultures of Southern Alaska: Rethinking Culture Area Boundaries.' *Arctic Anthropology* 29 (2):5–17.

———. 1998. 'Northern Northwest Coast Regional Overview.' *Arctic Anthropology* 35 (1):88–111. https://doi.org/10.2307/40316458.

———. 2004. 'The Status of Archaeology and Archaeological Practice in Southeast Alaska in Relation to the Larger Northwest Coast.' *Arctic Anthropology* 41 (2):177–96. https://doi.org/10.1353/arc.2011.0001.

———. 2011. *Northwest Coast: Archaeology as Deep History*. Washington, D.C.: Society for American Archaeology.

Moss, Madonna L., and Jon M. Erlandson. 1992. 'Forts, Refuge Rocks, and Defensive Sites: The Antiquity of Warfare along the North Pacific Coast of North America.' *Arctic Anthropology* 29 (2):73–90.

———. 1998a. 'A Comparative Chronology of Northwest Coast Fishing Features.' In *Hidden Dimensions: The Cultural Significance of Wetland Archaeology*, edited by Kathryn Bernick, 180–98. Vancouver: University of British Columbia Laboratory of Archaeology.

———. 1998b. 'Early Holocene Adaptations on the Southern Northwest Coast.' *Journal of California and Great Basin Anthropology* 20 (1):13–25.

———. 2001. 'The Archaeology of Obsidian Cove, Suemez Island, Southeast Alaska.' *Arctic Anthropology* 38 (1):27–47. https://doi.org/10.2307/40316538.

Moss, Madonna L., Jon M. Erlandson, R. Scott Byram, and Richard E. Hughes. 1996. 'The Irish Creek Site: Evidence for a Mid-Holocene Microblade Component on the Northern Northwest Coast.' *Canadian Journal of Archaeology* 20:75-. [PDF].

Muche, James. 1998. '[1981]. Site Location Factors. In the Realms of Gold: The Proceedings of the Tenth Conference on Underwater Archaeology, Edited by Wilburn A. Cockrell, Pp 137-139.' In *Maritime Archaeology: A Reader of Substantive and Theoretical Contributions*, edited by Lawrence Edward Babits and Hans Van Tilburg, 253–55. New York: Plenum Press.

Muckelroy, Keith. 1978. *Maritime Archaeology*. Cambridge: Cambridge University Press.

Muckle, Bob, and Alisha Gauvreau. 2017. 'Populating the Pacific Northwest.' *Anthropology News* 58 (2):148–51. https://doi.org/10.1186/s12905-016-0299-1.

Mulligan, Connie J., and Andrew Kitchen. 2013. 'Three-Stage Colonization Model for the Peopling of the Americas.' In *Paleoamerican Odyssey*, edited by Kelly E. Graf, Caroline V. Ketron, and Michael R. Waters, 171–82. Bryan: Center for the Study of the First Americans, Department of Anthropology, Texas A&M University.

Nance, Jack D. 1986. 'Reliability, Validity, and Quantitative Methods in Archaeology.'

Navy, US. 1998. 'US Navy Diving Manual, Chapter 1. Best Publishing Company, San Pedro, CA. 1985. Reprinted.' In *Maritime Archaeology: A Reader of Substantive and Theoretical Contributions*, edited by Babits and Tilburg, 343–54. New York: Plenum Press.

Nixon, F. Chantel, John H. England, Patrick Lajeunesse, and Michelle A. Hanson. 2014. 'Deciphering Patterns of Postglacial Sea Level at the Junction of the Laurentide and Innuitian Ice Sheets, Western Canadian High Arctic.' *Quaternary Science Reviews* 91. Elsevier Ltd:165–83. https://doi.org/10.1016/j.quascirev.2013.07.005.

NOAA. n.d. 'National Data Buoy Center.' https://www.ndbc.noaa.gov/.

Nobles, Gary. 2013. 'Spatial Analysis.' In *A Matter of Life and Death at Mienakker (the Netherlands): Late Neolithic Behavioural Variability in a Dynamic Landscape*, edited by J.P. Kleijne, O. Brinkkemper, R.C.G.M. Lauwerier, B.I. Smit, and E.M. Theunissen, 45:185–279.

Noort, Robert Van de. 2003. 'An Ancient Seascape: The Social Context of Seafaring in the Early Bronze Age.' *World Archaeology* 35 (3):404–15. https://doi.org/10.1080/0043824042000185793.

Oberg, K. 1973. *The Social Economy of the Tlingit Indians*. Seattle: University of Washington Press.

O'Claire, Rita, Robert H. Armstrong, and Richard Carstensen. 1992. *The Nature of Southeast Alaska*. Anchorage: Alaska Northwest Books.

Parker, A. J. 1999. 'A Maritime Cultural Landscape: The Port of Bristol in the Middle Ages.' *International Journal of Nautical Archaeology* 28 (4):323–42.

Patterson, Thomas C. 2008. 'The History of Landscape Archaeology in the Americas.' In *Handbook of Landscape Archaeology*, edited by Bruno David and Julian Thomas, 77–84. Walnut Creek, CA: Left Coast Press Inc.

Pearson, Charles E., David B. Kelly, Richard A. Weinstein, and Sherwood M. Gagliano. 1986. 'Archaeological Investigations on the Outer Continental Shelf: A Study within the Sabine River Valley, Offshore Louisiana and Texas.' Reston.

Peeters, Hans. 2005. 'The Forager's Pendulum: Mesolithic-Neolithic Landscape Dynamics, Land-Use Variability and the Spatio-Temporal Resolution of Predictive Models in Archaeological Heritage Management.' In *Predictive Modeling for Archaeological Heritage Management: A Research Agenda*, edited by Hans Kamermans and Martijn van Leusen, 149–68.

Amersfoort: Rijksdienst voor het Oudheidkundig Bodemonderzoek.

Peltier, W. R. 2004. 'Global Glacial Isostasy and the Surface of the Ice Age Earth: The ICE-5G (VM2) Model and GRACE.' *Annual Review of Earth and Planetary Sciences* 32:111–49.

Peltier, W. R., and Richard G. Fairbanks. 2006. 'Global Glacial Ice Volume and Last Glacial Maximum Duration from an Extended Barbados Sea Level Record.' *Quaternary Science Reviews* 25 (23–24):3322–37. https://doi.org/10.1016/j.quascirev.2006.04.010.

Peterson, Curtis. 1974. 'Data from Underwater Archaeological Sites.' In *Underwater Archaeology in the National Park Service: A Model for the Management of Submerged Cultural Resources*, edited by Daniel J. Leniham, 62–65. Washington, D.C.: Department of the Interior, US National Park Service.

PLOS. 2018. '13,000-Year-Old Human Footprints Found of Canada's Pacific Coast: New Evidence of Human Population Living on the West Coast of Canada at the End of the Last Ice Age.' *Science Daily*, March 2018. www.sciencedaily.com/releases/2018/03/180328143306.

Potter, Ben A., Joshua D. Reuther, Vance T. Holliday, Charles E. Holmes, D. Shane Miller, and Nicholas Schmuck. 2017. 'Early Colonization of Beringia and Northern North America: Chronology, Routes, and Adaptive Strategies.' *Quaternary International* 444. Elsevier Ltd:36–55. https://doi.org/10.1016/j.quaint.2017.02.034.

Potter, Ben A., James F. Baichtal, Alwynne B. Beaudoin, Lars Fehren-schmitz, C. Vance Haynes, Vance T. Holliday, Charles E. Holmes, et al. 2018. 'Current Evidence Allows Multiple Models for the Peopling of the Americas.' *Science Advances* 4 (August):eaat5473.

Purdy, B.A. 1991. *The Art and Archaeology of Florida's Wetlands*. Boca Raton: CRC Press.

Raab, L. Mark, and Albert C. Goodyear. 1984. 'Middle-Range Theory in Archaeology: A Critical Review of Origins and Applications.' *American Antiquity* 49 (2):255–68.

Raab, L. Mark, and Andrew Yatsko. 2009. 'California Galapagos.' In *California Maritime Archaeology: A San Clemete Island Perspective*, edited by L Mark Raab, Jim Cassidy, Andrew Yatsko, and William J. Howard, 3–22. New York: Altamira Press.

Raghavan, Maanasa, Matthias Steinrücken, Kelley Harris, Stephan Schiffels, Simon Rasmussen, Michael DeGiorgio, Anders Albrechtsen, et al. 2015. 'Genomic Evidence for the Pleistocene and Recent Population History of Native Americans.' *Science* 349 (6250). https://doi.org/10.1126/science.aab3884.

Ramenofsky, Ann F., and Anastasia Steffen. 1988. 'Units as Tools of Measurement.' In *Unit Issues in Archaeology: Measuring Time, Space, and Material*, edited by James M. Skibo, 3–17. Salt Lake City: University of Utah Press.

Ramsey, Carolyn L., Paul A. Griffiths, Daryl W. Fedje, Rebecca J. Wigen, and Quentin Mackie. 2004. 'Preliminary Investigation of a Late Wisconsinan Fauna from K1 Cave, Queen Charlotte Islands (Haida Gwaii), Canada.' *Quaternary Research* 62 (1):105–9. https://doi.org/10.1016/j.yqres.2004.05.003.

Reimchen, Tom, and Ashley Byun. 2005. 'The Evolution of Endemic Species in Haida Gwaii.' In *Haida Gwaii: Human History and Environment from the Time of the Loon to the Time of the Iron People*, edited by Daryl W. Fedje and Rolf W. Mathewes, 77–95. Vancouver: University of British Columbia Press.

Reimer, P. J., M. G. L. Baillie, E. Bard, A. Bayliss, J. W. Beck, P. G. Blackwell, C. Bronk Ramsey, et al. 2009. 'Intcal09 and Marine09 Radiocarbon Age Calibration Curves, 0–50,000 Years Cal BP.' *Radiocarbon* 51 (4):1111–50.

Rick, Torben C., Jon M. Erlandson, René L. Vellanoweth, and Todd J. Braje. 2005. 'From Pleistocene Mariners to Complex Hunter-Gatherers: The Archaeology of the California Channel Islands.' *Journal of World Prehistory* 19 (3):169–228. https://doi.org/10.1007/s10963-006-9004-x.

Ritter, Dale F., R. Craig Kochel, and Jerry R. Miller. 2006. *Process Geomorphology*. 4th ed. Long Grove, IL: Waveland Press, Inc.

Rossignol, Jacqueline, and LuAnn Wandsnider. 1992a. *Concepts, Methods, and Theory Building: A Landscape Approach. Space, Time, and Archaeological Landscape*. New York: Plenum Press.

———. 1992b. 'Theory Building A Landscape Approach.' In *Space, Time, and Archaeological Landscape*, edited by Jacqueline Rossignol and L Wandsnider, 3–19. New York: Plenum Press.

Ruppe, R. J. 1988. 'The Location and Assessment of Underwater Archaeological Sites.' In *Wet Site Archaeology*, edited by B.A. Purdy, 55–68. Telford: Telford Press.

Saile, Thomas, and Carsten Lorz. 2005. 'An Essay on Spatial Resolution in Predicitive Modelling: Pitfalls and Challenges.' Edited by Hans Kamermans and Martijn P. van Leusen. *Predictive Modeling for Archaeological Heritage Management: A Research Agenda*. Amersfoort: Colophon, 139–48.

Sandweiss, Daniel H., Heather McInnis, Richard L. Burger, Asunción Cano, Bernardino Ojeda, Rolando Paredes, María Del Carmen Sandweiss, and Michael D. Glascock. 1998. 'Quebrada Jaguay: Early South American Maritime Adaptations.' *Science* 281 (5384):1830–32. https://doi.org/10.1126/science.281.5384.1830.

Sawyer, Yadéeh E., Melanie J. Flamme, Thomas S. Jung, Stephen O. MacDonald, and Joseph A. Cook.

2017. 'Diversification of Deermice (Rodentia: Genus Peromyscus) at Their North-Western Range Limit: Genetic Consequences of Refugial and Island Isolation.' *Journal of Biogeography* 44 (7):1572–85. https://doi.org/10.1111/jbi.12995.

Schiffer, Michael Brian. 1972. 'Archaeological Context and Systemic Context.' *American Antiquity* 37 (02):156–65. https://doi.org/10.2307/278203.

———. 1988. 'The Structure of Archaeological Theory.' *American Antiquity* 53 (3):461–85.

Sebastian, Lynne, and W. James Judge. 1988. 'Predicting the Past: Correlation, Explanation, and the Use of Archaeological Models.' In *Quantifying the Present and Predicting the Past: Theory, Method, and Application of Archaeological Predictive Modeling*, edited by W. James Judge and Lynne Sebastian, 1–18. Denver: US Department of the Interior, Bureau of Land Management.

Shafer, Aaron B. A., Steeve D. Côté, and David W. Coltman. 2011. 'Hot Spots of Genetic Diversity Descended from Multiple Pleistocene Refugia in an Alpine Ungulate.' *Evolution* 65 (1):125–38. https://doi.org/10.1111/j.1558-5646.2010.01109.x.

Shafer, Aaron B. A., Catherine I. Cullingham, Steeve D. Côté, and David W. Coltman. 2010. 'Of Glaciers and Refugia: A Decade of Study Sheds New Light on the Phylogeography of Northwestern North America.' *Molecular Ecology* 19 (21):4589–4621. https://doi.org/10.1111/j.1365-294X.2010.04828.x.

Shennan, Stephen. 2002. *Genes, Memes and Human History: Darwinian Archaeology and Cultural Evolution*. London: Thames and Hudson.

Shugar, Dan H., Ian J. Walker, Olav B. Lian, Jordan B. R. Eamer, Christina Neudorf, Duncan McLaren, and Daryl W. Fedje. 2014. 'Post-Glacial Sea-Level Change along the Pacific Coast of North America.' *Quaternary Science Reviews* 97. Elsevier Ltd:170–92. https://doi.org/10.1016/j.quascirev.2014.05.022.

Smith, Jane L. 2011. 'An Update of Intertidal Fishing Structures in Southeast Alaska.' *Alaskan Journal of Anthropology* 9 (1):1–26.

Stanford, Dennis J., and Bruce A. Bradley. 2012. *Across Atlantic Ice: The Origin of America's Clovis Culture*. Berkeley: University of California Press.

Stanford, Dennis J., Darrin Lowery, Margaret A Jodry, Bruce Bradley, Marvin Kay, Thomas W. Stafford, and Robert J. Speakman. 2014. 'New Evidence for a Possible Paleolithic Occupation of the Eastern North American Continental Shelf at the Last Glacial Maximum.' In *Prehistoric Archaeology on the Continental Shelf: A Global Review*, edited by Amanda Evans, Joseph C. Flatman, and Nicholas C. Flemming, 73–93. New York: Springer. https://doi.org/10.1007/978-1-4614-9635-9.

State of Alaska, Department of Natural Resources. n.d. 'Alaska State Geo-Spatial Data Clearinghouse.' http://www.asgdc.state.ak.us/.

Stein, Julie K. 1992. 'Interpreting Stratification of a Shell Midden.' In *Deciphering a Shell Midden*, edited by Julie K. Stein, 71–93. San Diego: Academic Press.

Stern, Nicola. 2008. 'Stratigraphy, Depositional Environments, and Paleolandscape, Reconstruction in Landscape Archaeology.' In *Handbook of Landscape Archaeology*, edited by Bruno David and Julian Thomas, 365–78. Walnut Creek, CA: Left Coast Press.

Sturt, Fraser, Nicholas C. Flemming, D. Carabias, H. Jöns, and John Adams. 2018. 'The next Frontiers in Research on Submerged Prehistoric Sites and Landscapes on the Continental Shelf.' *Proceedings of the Geologists' Association*, no. April. The Geologists' Association. https://doi.org/10.1016/j.pgeola.2018.04.008.

Sullivan, Alan P. III, and Michael Brian Schiffer. 1978. 'A Critical Examination of SARG.' In *Investigations of the Southwest Anthropological Research Group: The Proceedings of the 1976 Conference*, edited by Robert C. Euler and George J. Gunerman, 168–75. Flagstaff: Museum of Northern Arizona.

Suttles, Wayne. 1990a. 'Environment.' In *Handbook of North American Indians, Volume 7: Northwest Coast*, edited by Wayne Suttles and William C. Sturtevant, 16–29. Washington, D.C.: Smithsonian Institution.

———. 1990b. 'Introduction.' In *Handbook of North American Indians. Volume 7: Northwest Coast*, edited by Wayne Suttles and William C. Sturtevant, 1–15. Washington, D.C.: Smithsonian Institution.

Swanton, John R. 1905. 'Contributions to the Ethnology of the Haida.' In *The Jesup North Pacific Expedition. Memoir of the American Museum of Natural History Volume V*, edited by Franz Boas, 1–300. New York: American Museum of Natural History.

Thompson, Laurence C., and M. Dale Kinkade. 1990. 'Languages.' In *Handbook of North American Indians, Volume 7: Northwest Coast*, edited by Wayne Suttles and William C. Sturtevante, 30–51. Washington, D.C.: Smithsonian Institution Press.

Tilley, Christopher. 1994. *Phenomenology of Landscape: Place, Paths, and Monuments*. Oxford: Berg.

———. 2004. *The Materiality of Stone: Explorations in Landscape Phenomenology*. Oxford: Berg.

———. 2008. 'Phenomenological Approaches to Landscape Archaeology.' In *Handbook of Landscape Archaeology*, edited by Bruno David and Julian Thomas, 271–76. Walnut Creek, CA: Left Coast Press Inc.

Trigger, Bruce G. 1971. 'Archaeology and Ecology.' *World Archaeology* 2 (3):321–36.

Tschauner, Hartmut. 1996. 'Middle-Range Theory, Behavioral Archaeology, and Postempiricist Philosophy of Science in Archaeology.' *Journal of Archaeological Method and Theory* 3 (1):1–30.

Verhagen, Philip. 2008. 'Testing Archaeological Predictive Models: A Rough Guide.' In *Layers of Perception: Proceedings of the 25th International Conference on Computer Applications and Quantitative Methods in Archaeology (CAA) Berlin, Germany, April 2-6, 2007*, edited by Posluschny Axel, Karsten Lambers, and Irmela Herzog, 285–91. Frankfurt: Romisch-Germanische Kommission des Deutschen Archaeolgischen Istituts.

Verhagen, Philip, and Thomas G. Whitley. 2012. 'Integrating Archaeological Theory and Predictive Modeling: A Live Report from the Scene.' *Journal of Archaeological Method and Theory* 19 (1):49–100. https://doi.org/10.1007/s10816-011-9102-7.

Wandsnider, LuAnn. 1992a. 'Archaeological Landscape Studies.' In *Space, Time, and Archaeological Landscape*, edited by Jacqueline Rossignol and LuAnn Wandsnider, 285–92. New York: Plenum Press.

———. 1992b. 'The Spatial Dimension of Time.' In *Space, Time and Archaeological Landscapes*, edited by Jacqueline Rossignol and LuAnn Wandsnider, 257–82. New York: Plenum Press. https://doi.org/10.4324/9780203109595.

Wang, Yue, Peter D. Heintzman, Lee Newsom, Nancy H. Bigelow, Matthew J. Wooller, Beth Shapiro, and John W. Williams. 2017. 'The Southern Coastal Beringian Land Bridge: Cryptic Refugium or Pseudorefugium for Woody Plants during the Last Glacial Maximum?' *Journal of Biogeography* 44 (7):1559–71. https://doi.org/10.1111/jbi.13010.

Warner, Barry G. 1984. 'Late Quaternary Paleoecology of Eastern Graham Island, Queen Charlotte Islands, British Columbia, Canada.' Simon Fraser University.

Watson, Patty Jo. 1983. 'Method and Theory in Shipwreck Archaeology.' In *Shipwreck Anthropology*, edited by Richard A Gould, 23–36. Albuquerque: University of New Mexico Press.

Watts, Jack, Brian Fulfrost, and Jon M. Erlandson. 2011. 'Searching for Santarosea: Surveying Submerged Landscapes for Evidence of Paleocoastal Habitation Off California's Northern Channel Islands.' In *The Archaeology of Maritime Landscapes*, edited by Ben Ford, 11–26. New York: Springer.

Weckworth, Byron V., Natalie G. Dawson, Sandra L. Talbot, Melanie J. Flamme, and Joseph A. Cook. 2011. 'Going Coastal: Shared Evolutionary History between Coastal British Columbia and Southeast Alaska Wolves (Canis Lupus).' *PLoS ONE* 6 (5). https://doi.org/10.1371/journal.pone.0019582.

Westerdahl, Christer. 1992. 'The Maritime Cultural Landscape.' *The International Journal of Nautical Archaeology* 21 (1):5–14.

Westley, Kieran, Trevor Bell, Ruth Plets, and Rory Quinn. 2011. 'Investigating Submerged Archaeological Landscapes: A Research Strategy Illustrated with Case Studies from Ireland and Newfoundland, Canada.' In *Submerged Prehistory*, edited by Johnathan Benjamin, Clive Bonsall, Catriona Pickard, and Anders Fischer, 129–44. Oxford: Oxbow Books.

Wheat, Amber D. 2012. 'Survey of Professional Opinions Regarding the Peopling of the Americas.' *SAA Archaeological Record* 12 (2):10–15.

Wheatley, David. 1993. 'Going over Old Ground: GIS, Archaeological Theory and the Act of Perception.' In *Computing the Past: Computer Applications and Quantitative Methods in Archaeology (CAA)*, edited by Anderson, Madsen, and Scollar, 133–38. Arahus: Aarhus University Press.

Wheatley, David, and Mark Gillings. 2002. *Spatial Technology and Archaeology: The Archaeological Applications of GIS*. London: Taylor and Francis.

Whitley, Thomas G., and Gwendolyn Burns. 2008. 'Conditional GIS Surfaces and Their Potential for Archaeological Predictive Modelling.' *Layers of Perception: Proceedings of the 35th International Conference on Computer Applications and Quantitative Methods in Archaeology (CAA), Berlin, Germany, April 2-6, 2007*, 292–98.

Wickert, Andrew D., Kelly Rose Bale Monteleone, and Robert S. Anderson. 2013. 'Reconstructing the Paleogeography of Beringia.' In *Poster Presented at the Paleoamerican Odyssey*. Santa Fe.

Wigen, Rebecca J. 2005. 'History of the Vertebrate Fauna in Haida Gwaii.' In *Haida Gwaii: Human History and Environment from the Time of the Loon to the Time of the Iron People*, edited by Daryl W. Fedje and Rolf W. Mathewes, 96–115. Vancouver: University of British Columbia Press.

Willey, Gorden R. 1953. *Prehistoric Settlement Patterns in the Viru Valley, Peru*. Washington, D.C.: Smithsonian Institution, Bureau of American Ethnology Bulletin 155.

Willey, Gorden R., and P. Phillips. 1958. *Method and Theory in American Archaeology*. Chicago: University of Chicago Press.

Williams, Mark R., and E. James Dixon. 2014. 'Labourchere Bay Final Report.' Juneau.

Wilson, Michael C. 1996. 'Late Quaternary Vertebrates and the Opening of the Ice-Free Corridor, with Speicial Reference to the Genus Bison.' *Quaternary International* 32:97–105.

Winterhalder, Bruce. 1981. 'Optimal Foraging Strategies and Hunter-Gatherer Research in Anthropology: Theories and Models.' In *Hunter-Gatherer Foraging Strategies*, edited by Bruce Winterhalder and E. A. Smith, 66–98. Chicago: University of Chicago Press.

Winthrop, Robert H. 1991. *Dictionary of Concepts in Cultural Anthropology*. New York: Greenwood Press.

Wissler, Clark. 1914. 'Material Cultures of the North American Indians.' *American Anthropologist* 16 (3):447–505.

———. 1917. *The American Indian: An Introduction of the Anthropology of the New World*. New York: Douglas C. McMurtrie.

Worl, Rosita. 1990. 'History of Southeast Alaska since 1867.' In *Handbook of North American Indians, Volume 7: Northwest Coast*, edited by Wayne Suttles and William C. Sturtevant, 149–58. Washington, D.C.: Smithsonian Institution.

Wright, John, Angela Colling, and Dave Park. 2005. *Waves, Tides and Shallow-Water Processes*. 2nd ed. Oxford: Butterworth-Heinemann.

Wygal, Brian T. 2018. 'The Peopling of Eastern Beringia and Its Archaeological Complexities.' *Quaternary International* 466. Elsevier Ltd:284–98. https://doi.org/10.1016/j.quaint.2016.09.024.

www.ingramcontent.com/pod-product-compliance
Ingram Content Group UK Ltd.
Pitfield, Milton Keynes, MK11 3LW, UK
UKHW061332200426
11947UKWH00044B/2013